AF335560

Fluid Mechanics and Dynamics of Multi-Valve Engines

Organizing Committee

C Brockley-Blatt
Ricardo Consulting Engineers

R Stone
University of Oxford

F Wallace
University of Bath

P Williams
University College, London (UCL)

W A Woods
Consultant

Papers presented at a one-day seminar *Fluid Mechanics and Dynamics of Multi-Valve Engines* held at University College, London, UK, on 9 June 1999.

IMechE Seminar Publication

Fluid Mechanics and Dynamics of Multi-Valve Engines

Jointly organized by the Energy Transfer and Thermofluid Mechanics Group, and the Combustion Engines Group of the Institution of Mechanical Engineers (IMechE)

Co-sponsored by :
Automobile Division (IMechE)
UNICEG
Cosworth Technology

IMechE Seminar Publication 1999–15

Professional Engineering Publishing

Published by Professional Engineering Publishing Limited for the Institution of Mechanical Engineers, Bury St Edmunds and London, UK.

ISSN 1357–9193
ISBN 1 86058 197 8

A CIP catalogue record for this book is available from the British Library.

Printed by Cromwell Press, Trowbridge, Wiltshire, UK.

Contents

Related Titles of Interest

Title	Editor/Author	ISBN
IMechE Engineers' Data Book	C Matthews	1 86058 175 7
Design Techniques for Engine Manifolds	D E Winterbone and R J Pearson	1 86058 179 X
An Introductory Guide to Industrial Flow	R Baker	0 85298 983 0
An Introductory Guide to Flow Measurement	R Baker	0 85298 670 X
Power Transmission and Motion Control (PTMC '98)	Edited by C R Burrows and K A Edge	1 86058 134 X
Power Transmission and Motion Control (PTMC '99)	Edited by C R Burrows and K A Edge	1 86058 205 2
CFD in Fluid Machinery Design	IMechE Seminar Publication	1 86058 165 X

For the full range of titles published by Professional Engineering Publishing contact:

Sales Department
Professional Engineering Publishing Limited
Northgate Avenue
Bury St Edmunds
Suffolk
IP32 6BW
UK

Tel: +44 (0)1284 724384
Fax: +44 (0)1284 718692

The use of multiple poppet valves in reciprocating internal combustion engines

R PEARSON, J MARTIN, and **I POSTLETHWAITE**
Lotus Engineering, Hethel, UK

ABSTRACT

The historical development of multiple poppet-valve configurations for reciprocating engines is briefly reviewed. Four-valve cylinder heads with pent-roof combustion chambers have been in use for almost 90 years. This technology, with somewhat more sophisticated intake flow control, is now used in the majority of modern engines for automotive applications.

The paper discusses the importance of valve area and the use of large-scale structures established during the induction period in order to manipulate the engine torque characteristic. The design of the inlet ports for both gasoline and diesel engines requires a compromise between maximizing the flow area available and the production of swirl motion in order to achieve acceptable combustion quality. In the gasoline case the use of shallow port angles to produce tumble swirl reduces the effective flow area through the valves. The Lotus Vortex Tumble Control System (VTCS) concept is presented. This approach offers a low cost mechanism for providing higher levels of flow area at maximum power with extended control over the combustion process.

1 Historical Background

The primary function of the valves in a reciprocating internal combustion engine is to admit and discharge the gas flowing through the engine. The size, number, and arrangement of the valves used to perform this task is, however, inextricably linked to the design of the engine combustion system. In modern engines the configuration of the valves not only enables the required amount of charge into the cylinders but establishes a carefully controlled flow field which is used to manipulate the combustion rate in the engine. This interdependence is an essential feature of current multi-valve spark-ignition engines, and is vital to the operation of the latest direct-injection spark-ignition and diesel engines.

As early as 1904 Clerk[1] had motored an engine with the valves held closed before ignition and found that the resulting combustion was very slow. Subsequent work by Ricardo[1,2] established the importance of the controlled use of in-cylinder turbulence in achieving adequate combustion rates across the engine speed range. Indeed, it was only by patenting his famous 'turbulent head' in 1920 that Ricardo was able to maintain the viability of the previously slow burning side-valve engines against the overhead-valve approach which was beginning to be adopted in production engines.

Because of its low cost the side-valve engine became the dominant automotive powerplant up to about 1950, and was still in production in some passenger car engines into the 1970's[1]. The more expensive overhead-valve engines, however, with their less restrictive flow paths into the cylinder, had superior volumetric efficiency to the side-valve engines. Inlet flow structures set up a natural mechanism for the production of turbulence, although this was not understood when the earliest overhead-valve engines were designed. These features, combined with the much shorter flame paths compared with side-valve engines, enabled overhead-valve engines to produce high BMEP values at elevated engine speeds and this type of design rapidly became the norm for high-performance engines.

1.1 Four-Valve Engines

Four-valve racing engines were built by Fiat as early as 1911[3,4]. Ernest Henry designed Peugeot racing engines which first appeared in 1912 having twin-overhead camshafts actuating two intake and two exhaust valves per cylinder, with an included angle of 60°; a pent-roof combustion chamber was used with centrally located spark plugs. In these engines the low bore / stroke ratios (0.5 to 0.6 up to the First World War) dictated the use of steep-sided pent-roofs to allow for the very large valve diameters thought to be required for high power output[4]. An additional feature allowing the use of larger valves was to make the combustion chambers over-hang the cylinder bores. Thus, in order to achieve compression ratios as low as even 5-6 to 1, some fairly severe wedge-topped or domed pistons were used, which had a deleterious effect on the structure of the in-cylinder fluid motion and produced undesirable 'orange-peel' shaped combustion chambers. Even so, compared with contemporary side-valve technology, these engines were a huge advance in terms of combustion chamber design and would have enabled increased compression ratios to be used for a given quality of fuel.

In 1919 Bentley manufactured a 3 litre racing engine with a single overhead-camshaft actuating four valves per cylinder, with a valve-included angle of 30 degrees - this was probably the first instance of a narrow-angle four-valve engine. As Lovell points out[4], however, the narrow valve-included-angle of this engine may have been dictated by the need to machine the valve seats via the cylinder bore because the head and block were integral. Borgeson[3] shows a section of a four-valve Miller 183 'straight-eight' engine produced in 1921 which has a pent-roof combustion chamber, a valve-included angle of 50 degrees, and a flat topped piston. The use of four twin-choke carburettors enabled each cylinder to have its own individual intake tract. This engine had a compression ratio of 7.5:1, an increase of 25 per cent over even the best two-valve overhead-valve designs of that year according to the data presented by Borgeson[3], and can , along with the Bentley engine, be thought of as a true predecessor of the modern four-valve engine.

High performance engines with four valves per cylinder were built by many other manufacturers, including Alfa-Romeo, Delage, Humber, Sunbeam, Frontenac, Ballot, Mercedes, Aston Martin, and Vauxhall, up to the mid-1920's. Rudd[5] states that most aircraft engines of the time had four valves per cylinder. Aircraft engines with four vertical valves per cylinder were made as early as 1917 by Napier (the Lion engine). This basic layout was commonplace in piston aircraft engines toward their apogee in the form of the Rolls Royce Griffon and Merlin of the 1930's and 40's. Some engines, such as the Rolls Royce Condor (1919) and Bristol Pegasus (1932), had inclined valves, although in the latter engine the two inlet valves were on opposite sides of the cylinder due to its radial configuration. The drive towards ever higher outputs and efficiencies in aircraft engines eventually led to the adoption of sleeve valves such as in the Bristol Centaur.

It seems, however, that the adoption of multiple overhead valves in early engines was not exclusively due to a desire for increased flow area. The two-principal reasons given by Lichty[6] for using multiple valves (more than two per cylinder) were: 1) lower valve weight; 2) increased valve seat area for better cooling - interestingly, it is precisely the same geometrical principles which endow multi-valve engines with superior flow area to two-valve engines. The introduction of sodium filled valves and stellite facing[7] made it possible to use two large valves reliably instead of four and the advent of supercharging from the mid 1920's onwards reduced the importance of valve flow area and intake system refinement[4]. Four-valve engines hence diminished in popularity in racing car engines from the mid 1920's until the late 1930's when they were briefly revived by Mercedes, Alfa Romeo, and Maserati, after which two-valve engines were dominant up to the mid 1960s.

In the motorcycle racing world Ricardo had introduced the four-valve concept in the 1921 single-cylinder 500 cm^3 Triumph Ricardo, with the valves arranged in parallel pairs at an included angle of 90 degrees. In 1928 Rudge produced four-valve heads with both parallel and radial inlet valve arrangements and radial exhaust valves, and in 1933 Excelsior produced a four valve 250 cm^3 engine with radial valves and separate inlet ports each served by their own carburettors[8]. The outstanding success of the two-valve single-cylinder 500 cm^3 racing Nortons, however, which were claimed to have produced 100 bhp/l in development versions prior to the Second World War[4], led European engine designers to largely abandon the four-valve layout in the early 1930's until Honda made very successful use of four-valve technology in its racing motorcycles in the 1960s. The Honda racing motorcycle engines had relatively wide valve included angles due to the use of air-cooled cylinders and ran at speeds up to 25,000 rev/min.[9], but by this time the narrow valve angle four-valve units with relatively flat topped pistons were emerging in Grand Prix car engines made by Weslake/BRM Gurney-Weslake and Cosworth[4] in the form now familiar in modern four-valve engines. This basic pent-roof design then became the norm in Grand Prix car engines.

The narrow valve angle four-valve system with pent-roof combustion chambers took longer to establish itself in production car vehicles. Triumph, with the Dolomite Sprint engine, Lotus, and Saab were early disciples of this valve configuration for production automotive vehicles but was not until the mid-1990's that the system was widely adopted in basic-specification mass produced engines. Amann[10] presents data showing the dramatic rise of the four-valve engine in the United States market. At the expense of the two-valve engine the market penetration of the four-valve engine rose from 1.6 per cent in 1986 to over 50 per cent within ten years. Amann[10] also shows that since 1990, when the market penetration of four-valve engines reached 25 per cent, the fleet-average specific output of the four-valve engine (measured in kW/litre) has exceeded that of two-valve units by an average of 36 per cent. This increase in specific output has enabled the 'downsizing' of powertrain units in the United States market[10], thereby giving increases in fuel economy.

1.2 Engines with More Than Four Valves Per Cylinder

Almost as soon as four-valve engines had appeared, Maybach[11] designed a five-valve engine for use in aircraft, airships, and marine applications. These engines had two inlet and three exhaust valves and was the most widely used German airship engine between 1915 and 1917[12]. Airship engines with six, seven, and eight valves per cylinder are mentioned by Rudd[5].

There have been several recent examples of engines which have more than four valves per cylinder. Arrangements with five-valves per cylinder (three inlet valves, two exhaust valves) have been used on high performance Yamaha motorcycles since the mid-1980's. Current production models using this technology are the 750 cm^3 YZF-R7 and the 1000 cm^3 YZF-R1; these four-cylinder engines are claimed to produce about 105 and 110 kW/litre at 11000 and 10000 rev/min respectively, giving specific torque figures of 96 Nm/litre at 9000 rev/min and 108 Nm/litre at 8500 rev/min. Whilst these figures are impressive there are examples of four valve engines which exceed the output values quoted above: notably the Suzuki GSX-R750 which produces a claimed 132 kW/litre and 110 Nm/litre, at respective engine speeds of 12300 and 10300 rev/min. Aprilia also use a single-cylinder engine with five-valves in their Pegaso and Moto 650 cm^3 motorcycles.

The 3.5 litre Ferrari F355, and its replacement, the 3.6 litre 360 Modena, use five-valve engines. In the latter model the specific outputs are 83 kW/litre at 8500 rev/min and 104 Nm/litre at 4750 rev/min. In this engine high specific outputs, with a large difference between the maximum torque speed and the maximum power speed, is achieved using a variable geometry intake system. Other high performance automobiles with five valves per cylinder have been produced in limited numbers: the Ferrari F50 and Bugatti EB110 GT both had V12 engines with this valve configuration. In the mid-1980's Maserati briefly produced a six-valve per cylinder unit.

Audi currently manufacture 1.8 litre in-line four, and 2.8 litre V6 engines which have five valves per cylinder, and have developed such engines for use in motorsport since 1984. Audi's rationale for adopting this technology in production engines is that it is necessary to compensate for the reduction of available flow area incurred by using small bore / stroke ratios which the company have adopted in order to reduce hydrocarbon emissions and improve part-load fuel consumption[12]. The same four-cylinder engines are also used in Volkswagen cars.

It is, of course, very difficult to compare production engines directly as it is the balance of a large number of design parameters which affects the output achieved by the engine. The shape of the torque curve is the limiting factor on the drivability/ridability of the vehicle in which the engine is installed and this must be considered carefully when analysing engine performance figures. Engines which might be most easily compared directly because they have the same swept volume, are naturally aspirated, and have a narrowly defined application, are those used in Formula 1 cars. Unfortunately reliable performance data for these engines is not generally available. Formula 1 racing engines which have been equipped with five valves per cylinder include the 1989 Tickford development of the Judd CV V8, the 1989 Yamaha OX88 V8 and the 1989 Ferrari V12. All these were 3.5 litre engines and had three inlet and two exhaust valves. Ferrari developed this unit in several stages but reverted to a conventional four-valve configuration part-way through the 1993 season. A cylinder head from one version of this engine is shown in Fig. 1. It is of interest to note that Yamaha-designed five-valve cylinder-heads were tested by Cosworth in 1987/88 for F1 use and abandoned[13].

Fig.1. Cylinder head from Ferrari Formula 1 engine. (Courtesy Ferrari GeS)

Perhaps the most extreme example of a multi-valve design is an eight-valve arrangement developed by Honda[14,15] which was initially used on a Grand Prix racing motorcycle and recently featured on a high-performance limited-volume production motorcycle. The original engine was a 500 cm^3 V4 unit produced in 1978 to compete in the 500 cm^3 Road Racing World Championship. This competition was dominated then, as it currently is, by two-stroke engines. In order to achieve a similar power output to a two-stroke engine with the same swept volume, a four-stroke engine has to operate at approximately twice the engine rotational speed. Since the regulations limit the maximum number of cylinders to four, the only way to achieve the very high engine speeds required was to use an ultra-short stroke - this gives the engine a concomitantly large bore.

The approach chosen by Honda to achieve the short flame-travel paths required at such high engine speeds was to use an elliptical bore cross-section with two spark-plugs and four inlet and four exhaust valves per cylinder. After four years development the Grand Prix engine eventually attained an output of over 100 kW at around 20,000 rev/min - a specific output of 268 bhp/litre (this figure is now approached by Formula 1 racing engines using four-valve technology). A 0.75 litre endurance racing motorcycle engine and a 'production' engine (giving maximum power at 14,000 rev/min) of the same capacity were subsequently made using eight valve technology.

2 EFFECTIVE FLOW AREA

Livengood and Stanitz[16] attempted to formulate a parameter which characterises the 'average' Mach number of the gas flowing through the inlet valve by considering engines charging and discharging from reservoirs (i.e. without significant inlet pipes). Assuming that the flow of gas through the inlet valves into the engine cylinder is incompressible, and occurs between top-dead-centre (tdc) and bottom-dead-centre (bdc) of the induction stroke, enables a mean inlet gas velocity, $\bar{u}_v$, to be obtained from the relationship

$$\bar{u}_v = \frac{\bar{U}_p F_c}{\bar{C}_f F_v}, \qquad\qquad (1)$$

where $\bar{U}_p$ is the mean piston speed, F_c is the cylinder cross-sectional area, and F_v is the cross-sectional area of the valve port. $\bar{C}_f$ is the mean flow coefficient given by

$$\bar{C}_f = \frac{\bar{F}_{v_{eff}}}{F_v} \qquad\qquad (2)$$

where $\bar{F}_{v_{eff}}$ is the inlet valve mean effective area [m^2], defined as

$$\bar{F}_{v_{eff}} = \int_{\theta_{ivo}}^{\theta_{ivc}} C_{f_i} F_{v_i}(\theta)\, d\theta \frac{1}{(\theta_{ivc} - \theta_{ivo})} . \qquad (3)$$

This enables the Inlet Mach index, sometimes referred to (rather unjustly!) as the 'Taylor Mach index', or 'Gulp Factor', to be expressed as

$$Z = \frac{\bar{U}_p}{a} \frac{F_c}{\bar{F}_{v_{eff}}} = \frac{V_{swept}\, N}{30 a\, \bar{F}_{v_{eff}}} , \qquad\qquad (4)$$

where a is the speed of sound in the inlet gas, V_{swept} is the cylinder swept volume, and N is the engine speed in rev/min. Clearly the derivation of this parameter is based on the assumption that the volumetric efficiency on the cylinder is 100 %.

Plotting the variation of volumetric efficiency against Inlet Mach index gives a curve of the general form shown in Fig. 2. There is a gradual reduction in volumetric efficiency above values of Inlet Mach index of about 0.6 and this has lead to the widespread adoption of equation (4) in predicting the limiting engine speed for a given inlet valve mean effective area.

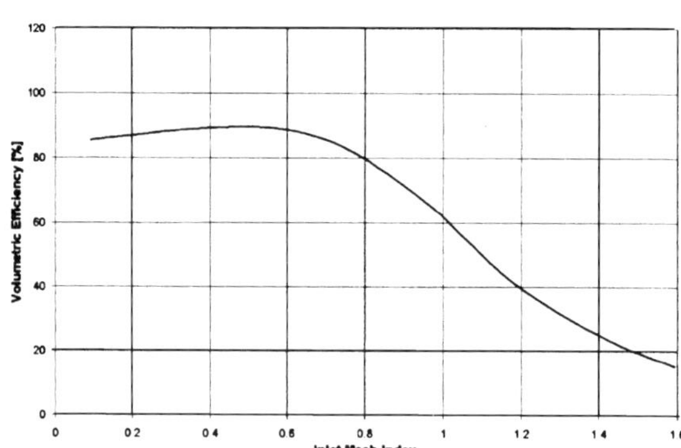

Fig.2 Variation of volumetric efficiency with Inlet Mach index

Equation (4), however, does not characterize the dependence of the limiting value of Inlet Mach index on the duration of the intake valve opening event and it has been found that volumetric efficiency does not always decrease when Z is in the range 0.5-0.6[17]. In order to represent the critical parameters more rigorously Fukutani and Watanabe[17] defined the Mean Inlet Mach number ($\overline{M}_v$) as

$$\overline{M}_v = \frac{\overline{u}_v}{a} = \frac{V_{swept}\,\eta_v}{a\,F'_{v_{eff}}\,100} \qquad (5)$$

where

$$F'_{v_{eff}} = \int_{t_{ivo}}^{t_{ivc}} C_{f_i} F_{v_i}(t)\,dt\,, \qquad (6)$$

and η_v is the volumetric efficiency in per cent. Note that the units of the parameter $F'_{v_{eff}}$ are $[m^2\ s]$, i.e. the parameter represents the time integral of inlet valve area. Substituting equation (3) in equation (6) gives

$$F'_{v_{eff}} = \overline{F}_{v_{eff}}\left(t_{ivc} - t_{ivo}\right) = \overline{F}_{v_{eff}}\frac{\left(\theta_{ivc} - \theta_{ivo}\right)}{6\,N}\,, \qquad (7)$$

from which it is clear that $F'_{v_{eff}}$ is a function of engine speed and valve timing. Combining equations (5) and (7) yields the Mean Inlet Mach number in the form

$$\overline{M}_v = \frac{V_{swept}\,\eta_v}{a\ 100}\frac{6\,N}{\overline{F}_{veff}\left(\theta_{ivc} - \theta_{ivo}\right)}\,. \qquad (8)$$

Fukutani and Watanabe[17] showed that plotting volumetric efficiency against Mean Inlet Mach number for engines without intake pipes produces a curve of the generic form shown in Fig. 3. It can be seen that Mean Inlet Mach number gives a well defined sudden reduction in the breathing efficiency of the engine at around $\overline{M}_v$=0.5. This dramatic reduction in volumetric efficiency is due to the value of the Mach number at the valve throat, based on the local speed of sound at the inlet valve, being unity, i.e. the flow being choked, for most of the period of the induction stroke. The sonic flow in the throat of the valve is caused by the ratio of the pressure in the cylinder to that in the inlet pipe falling below the critical condition given by

$$\frac{p_c}{p_i} \leq \left(\frac{2}{\kappa+1}\right)^{\frac{\kappa}{\kappa-1}}\,. \qquad (9)$$

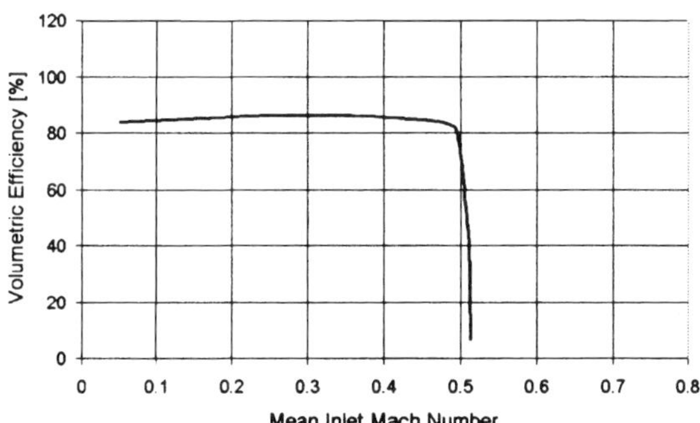

Fig.3. Variation of volumetric efficiency with Mean Inlet Mach number

For air, with κ=1.4, the critical ratio given by equation (9) is 0.528. Fukutani and Watanabe[17] show that when the Mean Inlet Mach number approaches 0.5 the inlet pressure ratio becomes critical during most of the inlet valve opening period, and the flow rate into the cylinder is then determined only by the inlet valve effective time area integral (equation (7)). Thus, in order to achieve adequate cylinder charging in high speed engines it is essential to maintain the value of $F'_{v_{eff}}$ by increasing the value of the inlet valve mean effective area, $\overline{F}_{v_{eff}}$ as N increases, thereby keeping the value of $\overline{M}_v$ below 0.5 .

Re-arranging equation (8), and setting $\overline{M}_v$ to 0.5, gives

$$\eta_v = \frac{50a}{6} \frac{\overline{F}_{v_{eff}} \left(\theta_{ivc} - \theta_{ivo}\right)}{V_{swept}} \frac{1}{N} = \frac{K}{N} , \qquad (10)$$

This equation shows that, once the flow at the valve is choked over most of the inlet period, volumetric efficiency is inversely proportional to engine speed for fixed inlet valve area and duration. Fig. 4 shows how the volumetric efficiency varies with engine speed for given K values. Obviously this relationship is not realistic over the engine speed range but it can be used to indicate the speed at which the volumetric efficiency of the engine, without manifolds fitted, will become limited by the valve area.

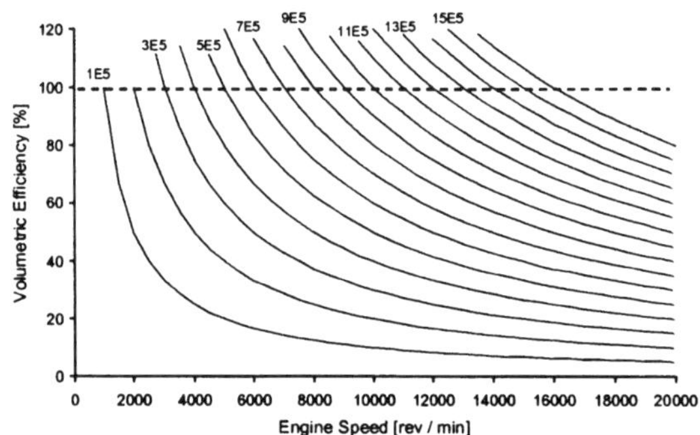

Fig.4. Variation of volumetric efficiency with engine speed for lines of constant K

It should be noted at this point that the variation of engine volumetric efficiency with Inlet Mach index and the Mean Inlet Mach number differs markedly from that shown in Figs. 2 and 3 when engines fitted with realistic intake manifolds are considered[18,19]. In this case there is significant pressure wave activity upstream of the intake valve and this modifies the pressure ratio across the valve. The pressure wave dynamics can be used to extend the valve opening period over which it is possible to achieve flow into the cylinder. Though the Mean Inlet Mach number remains something of a simplification of the actual process of cylinder charging, it can be used as a basic design parameter for engines when it is desired to consider the charging of the cylinder, independently of the intake system.

The basic geometric principles involved in maximizing mean inlet valve effective area are reviewed by Woods and Brown[20], and a comprehensive study, culminating in the production of five-valve engines for a motorcycle, was carried out by Aoi, Nomura, and Matsuzaka[21]. In this work an analysis similar to that of Woods and Brown was performed using the design criteria that the minimum clearance between valve seats was 2 mm for intake valves and 4 mm for exhaust valves; the ratio of the intake valve opening area is about 1.3 times bigger than that of the exhaust valve. Aoi et al[21] showed that calculated intake valve opening area, for a given bore diameter, of the five-valve configuration was bigger than that of four-, six-, and seven-valve designs. Four- and five-valve engines of the same bore (57mm) were then built and steady flow tests showed a superior mass flow rate through the cylinder of the five-valve engine across the lift range for a fixed pressure drop. The volumetric efficiency of the five-valve engine proved to be about 10 per cent higher than the four-valve engine in the high speed range (7000-13000 rev/min). For similar valve-spring design constraints the reduced equivalent weight and lower lift of the five-valve design increased the speed at which valve bounce occurred by about five per cent.

Work by Ohrnberger et al.[12] also shows higher values of calculated geometric area for a five-valve against a four-valve design. They also state that the large webs between the spark-plug and the valves, in the case of the five-valve design, enable effective water cooling around the spark-plug which gives increased knock-resistance. Additional advantages found by Audi are reduced valvetrain mass and camshaft drive torque compared with the four-valve engine.

Significant increases in valve area at both low and high valve lifts are claimed by Downing[22], who also examines the high sensitivity of five-valve systems to shrouding of the outer inlet valves by the cylinder walls and the interference that occurs between the flow from by the central inlet valve and the two outer valves. Downing shows that a reduction in the valve-head diameters actually increased the airflow at L/D ratios of 0.1 to 0.35, due to decreased masking and inter-valve flow interference. Improvements to the central inlet valve were also obtained by relieving the combustion chamber of this valve in order to allow more gas to leave the port to the sides and rear of the valve, thereby reducing interference with the two outer inlet valves.

Sykes[23] presents a figure comparing calculations of inlet valve area variation with bore size for 'typical' four- and five-valve designs. The calculations account for the effects of design features such as the outside diameters of the valve seat inserts and the minimum acceptable distances between them, but no theory is presented. The curves show that, for engines with bore diameters below about 80 mm, there is little to be gained by adopting the five-valve. This trend is in line with the fact that Yamaha presently manufacture high-performance motorcycle engines of 1000 and 750 cm^3 displacement having five valves per cylinder (the R1 and R7 motorcycles) but use a 600 cm^3 engine with four valves per cylinder in the R6 motorcycle.

In fact the five-valve system presented by Sykes[23] had a slightly lower inlet valve effective area than the four-valve system it replaced, a fact attributed to the inlet valve angle being increased over the four-valve engine to obtain more tumble swirl. Joyce et al.[24] report 'no functional advantage' of five-valve heads over four-valve units for the luxury passenger car engine which they developed, and Crouch et al.[25] found only one per cent improvement in the inlet valve effective area available from the use of five-valve technology in a gasoline engine.

It is apparent that there is some discrepancy between the increases in performance claimed by the five-valve systems of different manufacturers (and that achievable in practice!). Whilst this discrepancy may be due to detailed design differences between the various systems, any increase in flow area actually achieved may be more than off-set by necessary compromises in the design of the combustion chamber. A further factor which may affect the volumetric efficiency obtained on an engine, as opposed to the valve area measured on a steady flow bench, is the difference in the pressure wave propagation and reflection which can be induced by the complex port geometry of five-valve systems.

Studying the Honda eight-valve system, Kanazawa et al.[15] showed that the measured mean effective intake valve area obtained using the elliptical cylinder bore, with four inlet valves, was 29 per cent greater than that obtained from a 'corresponding' cylinder having a circular cross-section, equal in area to the elliptical cylinder, and two inlet valves. The elliptical cylinder was also shown to give a 12 per cent increase in mean effective intake valve area compared with two equivalent area circular cylinders, each with two inlet valves. Since, in this case, the number of valves and their diameters were the same the increase in flow area was ascribed to the reduced effects of interference of the incoming air flow with the cylinder walls experienced by the central two inlet valves in the elliptical cylinder.

3 IN-CYLINDER AIR MOTION

It was established by Ricardo[1] before 1920 that fluid motion within the cylinders of internal combustion engines is one of the most important factors controlling the combustion process. Studies on the use of axial and tumble swirl in order to reduce cyclic variability and combustion duration are reported in literature dating from 1939[26,27]. These concepts are essential to the successful operation of modern gasoline and diesel engines.

3.1 Air Motion in Gasoline Engines

Flame kernel development in gasoline engines is very sensitive to the turbulence field in the vicinity of the spark plug at the point of ignition. Cyclic variations in the flame kernel shape and heat transfer rates are caused by distortions in the velocity field locally around the spark plug and this causes significant variability in lean combustion. At fuel-air ratios close to stoichiometric a more important effect on the fully developed flame propagation is the stability, i.e. repeatability, of the bulk flow structure. The ratio of turbulent to laminar burning velocity is proportional to turbulence intensity which, in turn, is proportional to the average piston velocity. At a given engine speed the turbulence intensity in the cylinder is dependent on the large-scale flow structures established during the induction process, and their subsequent breakdown into turbulence as the piston rises in the compression stroke.

Modern automotive engines require high specific output levels combined with significant lean burn capability and EGR tolerance in order to meet performance, fuel economy, and emissions targets. Gasoline engines with multiple inlet valves offer large effective flow area which is desirable at high engines speeds in order to produce maximum power but at low engine speeds, and in particular at low loads, gas velocities, and hence turbulence levels, are low and this limits the propagation speed of the flame and EGR tolerance. These problems are generally now addressed by careful design of the inlet system and combustion chamber in order to produce net angular momentum about an axis normal to that of the cylinder. This controlled vortex motion, known as tumble, or barrel swirl, results from an uneven mass flow distribution over the intake valve[28,29], and breaks down during the second half of the compression stroke, releasing its bulk kinetic energy as turbulence[28-30]. The pent-roof combustion chamber with four valves per cylinder provides the ideal geometry to apply this technology and enables balance between flow area and tumble motion to be dictated by the angle of the inlet port to the horizontal[31]. There are, of course, NVH issues to be addressed when deciding upon a combustion chamber design. The main criterion in this respect is the maximum rate of change of cylinder pressure, which is manifested as the perceived level of combustion harshness.

3.2 Air Motion in Diesel Engines

In this Section the importance of air motion established in the induction process with respect to combustion in direct-injection (DI) diesel engines is discussed briefly. The mixing of the air and fuel in indirect-injection diesel engines takes places mainly in a pre-chamber into which air is forced through a narrow throat during the compression stroke, and is largely independent of the intake flow structure.

In small automotive DI diesel engines structured in-cylinder air motion is essential to achieving the mixing rates and spray penetration necessary for operating at high engine speeds with low particulate emissions. The simplest approach to providing this air motion is to use a masked, or shrouded, inlet valve which imparts a directional bias on the in-flowing air without the need for a complex intake port design. Since masked valves partially obstruct the area available for gas flow they are unsuitable when high specific outputs are required.

Engines intended for relatively high-speed applications (in the diesel context) require inlet ports which establish the required air motion in the cylinder. In modern engines the use helical inlet ports has been widely adopted. These ports provide structured axial swirl in the cylinder which is enhanced by the use of carefully designed piston bowls. The major effect of the intake process is to determine the large-scale structure of the gas motion in the combustion chamber at the inlet valve closing point. During the compression stroke the gas is pushed into the piston bowl, thereby increasing the swirl rate due to conservation of angular momentum. This swirl improves air utilization in small engines by deflecting the fuel sprays, thereby reducing spray impingement on the piston bowl which can limit the utility of high pressure injection systems by preventing sufficient air entrainment and mixing. Swirl can also be used to 'lean out' the pre-mixed burning phase in DI diesel engines, thereby limiting the NO_x emissions associated with the high cylinder pressures and temperatures produced by this part of the combustion process. One of the significant benefits of turbocharging DI diesel engines is the reduction in ignition delay due to the higher temperatures and pressures of the charge gas during injection, and the consequent reduction in pre-mixed burning which produces lower NO_x levels[26]. As with tumble motion in gasoline engines, swirl has the effect of stabilizing the gas flow within the cylinder and can reduce cyclic variation in the combustion process.

Work by Tindal et al.[27] has shown that the air motion induced by swirl ports is far from that of solid body rotation about the cylinder axis. Significant circulation in the vertical plane is produced by the gas leaving the valve, and this motion is augmented as the gas is deflected by the piston. There is also a significant amount of squish and reverse squish as the gas enters and leaves the piston bowl during the compression and expansion processes.

The production of strong axial swirl motion in the cylinder comes at the expense of reduced flow capacity through the helical port - this point will feature in the next Section of the paper.

4. FLOW AND COMBUSTION SYSTEM STRATEGIES IN PRACTICE

This section discusses the different port and valve strategies employed in the major engine types.

The wide range of valve and port designs used throughout the development of the reciprocating engine demonstrate great diversity and success in achieving the targets of the day. Present day engines exhibit no less diversity. Meeting and exceeding consumer expectations for performance, economy and refinement while successively reducing emissions has demanded extensive advancement in the understanding of flow and combustion phenomena.

To understand why different designs for the intake and combustion system are used it is important to consider the product's goals. The targets for performance, economy and emissions largely determine the balance between an outright drive for maximum performance, greater attention to the combustion quality and control, or attention to the spread, or 'elasticity' , of torque across the engine speed range. In order to demonstrate how flow and combustion strategies are driven by the design targets a range of well known engine types will be considered and the various design features which have been developed to meet today's high expectations will be discussed.

4.1 Contemporary Formula One Racing Engine

These engines are obviously designed to achieve very high levels of power, particularly in specific terms. To achieve high power for a given engine capacity (in this case 3.0 litres), the design maximum power speed is very high (>15000rpm) and the maximum achievable volumetric efficiency is needed. Figure 5, illustrates the performance targets along with a representation of a typical port design of this type.

As the figure shows, given such high volumetric efficiency requirements, the engine design is driven by the need to maximise valve area, and reduce flow losses in the port and valves to a minimum. To this end, high performance port geometry generally exhibit very few corners or tight radii where flow separation and recirculation might occur.

The port approach angle relative to the cylinder axis is steep in order that the full geometric area of each valve is presented to the on-coming air. Clearly to approach the cylinder from a shallow angle would reduce to effective port throat flow area to the detriment of flow efficiency.

To maintain sufficient combustion rates commensurate with adequate burning of the trapped charge, in-cylinder turbulence rates must be high. Fortunately, the high engine speed and gas velocities mean that in practice in-cylinder turbulence is usually adequate to negate significant detonation difficulties, although designs maximising valve diameter and lift have in the past tended to suffer from poor piston crown form and it's interaction with the burning charge.

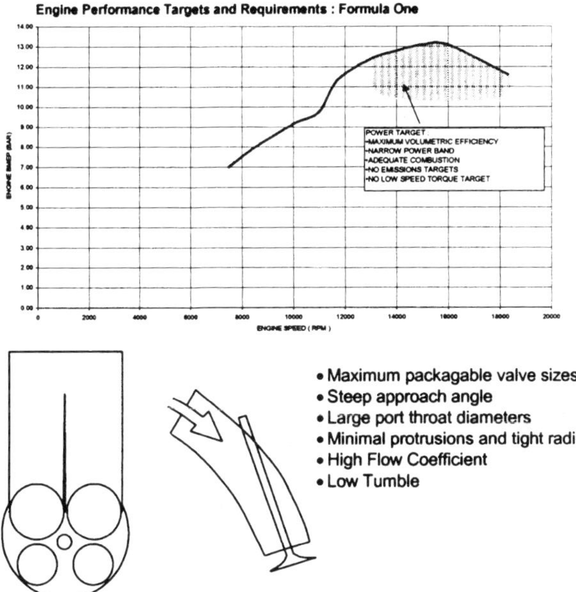

Fig.5. Formula One Engine Performance Targets and contemporary port and valve configuration

4.2 High Performance Motorcycle Engine
High performance motorbike engines achieve the highest levels of power per litre of any mass produced engine type. Typically these engines are capable of between 120 and 200 bhp/litre.

The flow and combustion strategies employed are quite similar to those of racing engines such as the Formula One engines described above. High levels of power are required, but in addition the consumer generally demands improved levels of torque in the middle of the engine speed range and improved drivability (typified by a smoother spread of torque across the whole speed range). These targets are represented in Figure 6 along with the major features of the port and valve design of engines in this class.

As the figure shows, although high flow efficiency is an important design criteria, the lower engine speeds of production machines and the need to achieve adequate mid-speed torque demands that the design be augmented to deliver increased levels of in-cylinder tumble. To achieve this the inlet port approach angle is relatively closer to the horizontal compared to Formula One practice and the port has a tighter bend radius just prior to the valve. The sharper deviation in the flow path causes the flow to separate from the port floor and hence more of the flow into the cylinder occurs across the spark plug side of the inlet valve. This flow induces a barrel 'tumble' motion within the cylinder which promotes high levels of turbulence. High levels of turbulence promote increased burn rates during combustion and permit higher maximum cylinder pressures at lower speeds, higher torque and improved performance across the engine speed range. Unlike the Formula One engines, the low speeds encountered during operation demand this strategy in order to alleviate knock which would otherwise limit low speed performance. The large valve sizes required to deliver high specific power would otherwise lead to low gas velocities and insufficient turbulence levels in this operating regime.

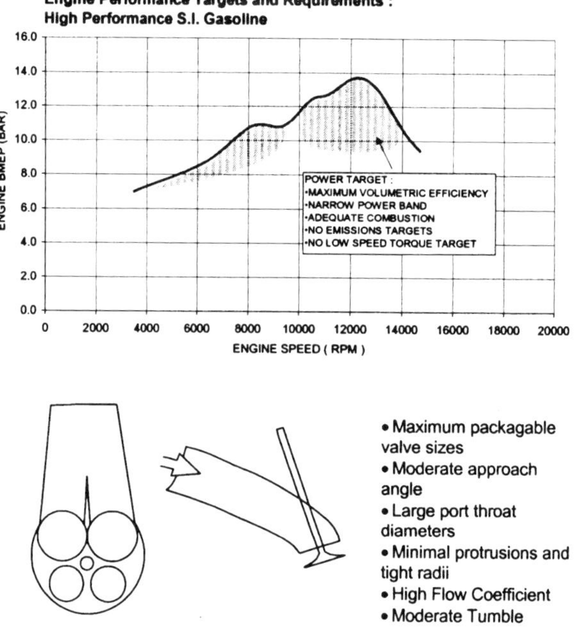

Fig.6. High Performance Motorbike Engine Targets and Port/Valve Geometry

4.3 Passenger Car S.I. Gasoline

Modern passenger cars achieve levels of specific power and torque on average 10-30% greater than similar vehicles of 10 years ago alone with significant gains in fuel economy. This is primarily due to the improvements achieved in flow and combustion performance as a result of the rapid transition to four-valve per cylinder technology in the early to mid 1990's.

The targets of a passenger car engine are quite different from the comparatively more straightforward racing and high performance motorbike engines. This arises from the greater range of operational targets which this type of engine is required to meet, both as a result of ever increasing performance and refinement demands from the consumer, government legislation on emissions and to reduce fuel consumption.

Figure 7 presents the targets for a typical contemporary passenger car type and demonstrates the need for both full load performance and part load economy and cleanliness. To meet these targets, designers have combined the benefits of efficient four valve per cylinder / central spark plug arrangements, with the potential for improved knock resistance and combustion control achieved through tumble augmented in-cylinder flow.

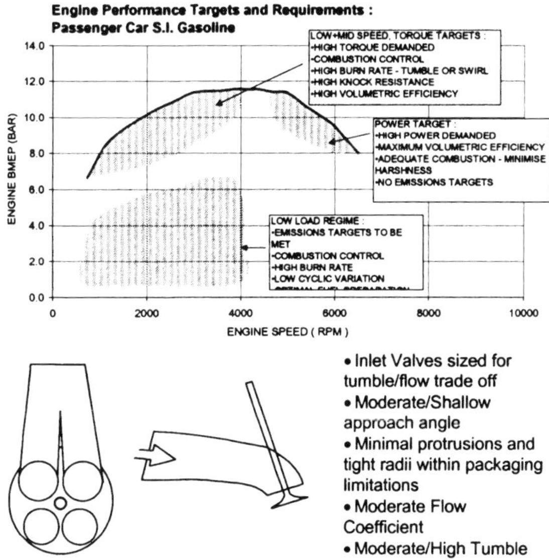

Fig.7. Passenger Car Gasoline SI Engine Targets and Port/Valve Geometry

The inlet valves are generally sized to within the limits of acceptable valve seat clearances bearing in mind thermal loads and manufacturing tolerances. The inlet port follows the progression from the motorbike design – as engine speed falls, the level of turbulence generated from gas velocity falls and the system needs to be highly augmented by a shallow port approach and sharp bend radii to provide high rates of tumble. The designer is forced to accept a trade-off between flow efficiency at high engine speed to achieve adequate power, and levels of tumble commensurate with good knock resistance and combustion control in order to deliver high low-speed torque, low emissions and good fuel consumption.

4.4 Passenger Car Turbocharged Diesel
Figure 8, below presents the targets and typical valve and port arrangement for a modern passenger car diesel engine. Passenger car diesel engines have, over the past few years converged on a swirl based combustion system in order to rapidly flow air across the finely atomised fuel plumes during the injection/combustion process. In order to minimise emissions and extend the engine's maximum speed and power capability this has necessitated a high rate of swirl in the cylinder. This is achieved by a port design, which generates a strong axial swirl by virtue of the asymmetric port arrangement, very shallow port approach angle, small inlet valve sizes and so-called 'helical' port throats.

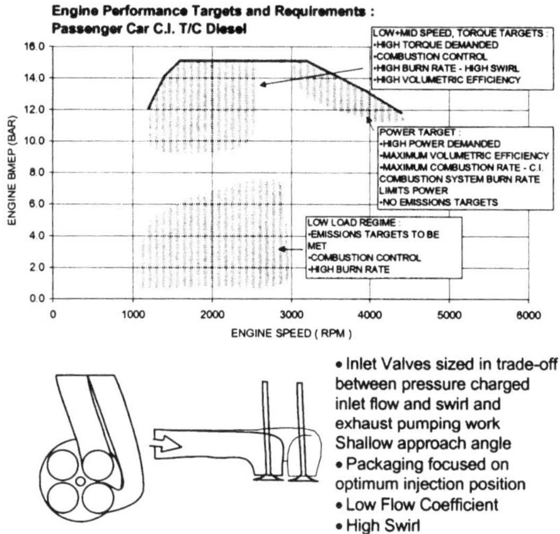

Fig.8. Passenger Car Diesel T/C Engine Targets and Port/Valve Geometry

These ensure that the flow enters the cylinder as a pair of carefully directed high velocity jets. This is effective in generating high axial swirl rates, thus supporting an efficient mixing of fuel and air giving relatively high burn rates (thereby increasing engine maximum speed). The port design tends, however, to reduce the volumetric efficiency of a naturally aspirated engine to the point at which the pumping losses increase greatly and the thermal efficiency is compromised. To overcome this obstacle and achieve parity of performance with S.I. engines, most mainstream B and C-class vehicles employ turbocharging on the powertrain. Turbocharging the engine provides high volumetric efficiency and benefits emissions by ensuring adequate excess air is present at all speeds and loads to obviate excess soot generation. Additionally, turbocharging raises the in-cylinder pressures and this further increases burn rate. Despite these benefits, at low loads, when boost is not sufficient, the momentum in the cylinder may be so low that it is necessary to de-activate one of the valves (increasing gas jet velocity on the remaining one), or augment, using valves or flaps, the aerodynamic characteristics of the port. This additional technology is expensive and imparts some pumping losses - ironically opposing one of the primary motivations for having adopted diesel in the first place.

Clearly the need to generate adequate combustion efficiency and power levels drives the port design. The modern diesel's inability to meet all targets via the porting and valve system alone is the primary reason why the cost of diesel engines remains high relative to their gasoline counterparts. The high price of the fuel injection system is a significant component of this cost but is necessary due to the inability of the fluid control system to atomise the fuel using aerodynamic shear alone.

4.5 High Performance Passenger Car S.I. Gasoline with Port De-Activation
In order to meet relatively aggressive targets for specific performance, torque and fuel consumption whilst producing low emissions, some Japanese and U.S. manufacturers adopted a strategy combining concepts from both spark ignition and diesel engine technology.

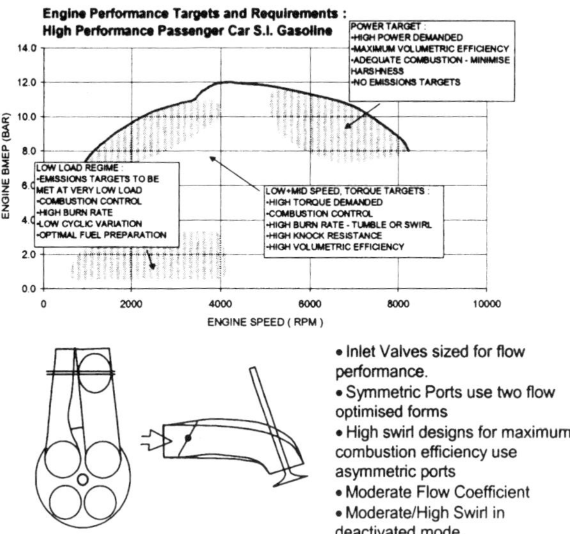

Fig.9. High Performance Passenger Car SI Gasoline with Port De-Activation

Figure 9 presents the use of port/valve de-activation matched to asymmetric porting, although some strategies use symmetric geometry.

The main driver for this approach was the need to achieve good volumetric efficiency at high engine speeds and loads. As the figure shows, this arrangement combines a high flow efficiency port connected to one valve (similar to motorbike/high performance practice) with a swirl port feeding the other valve. Although this arrangement may compromise the maximum mass flow rate of air through the engine, depending on whether the second port is a dedicated swirl port of is simply a standard port, the benefits of the system lie in it's ability to produce sufficiently high levels of axial swirl for the engine to operate lean at part load. Horie et al[34], show how this strategy can enable the engine to operate at air-fuel ratios of 22:1 at 1.6 bar load whilst producing a specific fuel consumption improvement of around 12% and a 70% reduction in engine out NO_x emissions (sufficient to meet Federal standard for NO_x over the drive-cycle). In the lean regime, the catalyst is operating essentially as an oxidation device whilst the cylinder temperatures (via charge dilution and part load operation) are low enough for low NO_x conversion efficiency to be inconsequential.

The engineer is forced to weigh up the trade-off in the potentially lower durability and increased system cost and complexity. This technology has also been applied in niche markets models where either the required tuning for performance or the need to deliver economy and emissions compliance in vehicles with high performance engines forces the adoption of artificial turbulence augmenting systems.

4.6 Passenger Car S.I. with GDI
In the past few years there has been an explosion of interest into the application of gasoline direct injection as a strategy for reduced fuel economy and emissions. This technology primarily focuses on the potential for burning gasoline in a locally stoichiometric air/fuel mixture zone stratified from the bulk of the gas in the cylinder (which may be unmixed air or

re-circulated exhaust gas, EGR) so that the engine can run significantly lean over large portions of it's operating regime.

To achieve this goal the aim of the engine designer is to maintain stratfication of the charge and surrounding gas as much as possible. Mitsubishi achieved this (in so much as the product is in the market) by developing a port and valve design matched to a deflector piston. Figure 10 presents this design and shows the engine operational requirements across the engine map.

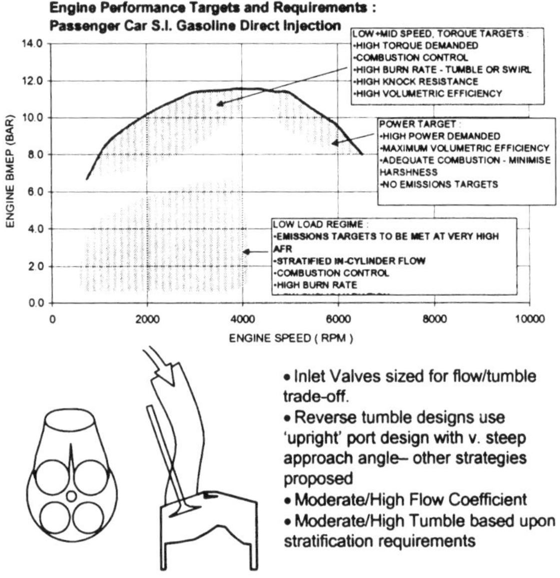

Fig.10. GDI Performance Targets and Port/Cylinder Geometry

The inlet ports impart a flow structure described as 'reverse tumble', which provides high flow efficiency for performance and can be tuned to produce a strongly stratified fuel-air mixture when required. Arguably this design goal could be achieved by careful development of the standard automotive tumble port in it's current form – the packaging of the upright reverse tumble port is a major issue when bonnet lines are being forced ever lower by crash, styling and aerodynamic demands. The validity of the port design in it's application is not the main issue. More importantly it highlights the ever greater emphasis being placed on the design and development of fluid control technologies to meet the increasing performance, economy and emissions targets of the future.

4.7 Systems for Contemporary Port Design and Development
As has been shown, a wide range of port-valve systems are currently in use, all driven by targets for performance, economy and emissions (and some secondary issues such as NVH not discussed here).

To further optimise current systems and develop new strategies to enhance control and performance, a range of approaches have been used and new techniques continue to emerge. These are listed for completeness; detailed discussion is omitted for brevity:

- Steady Flow Rig Testing – Laboratory testing of port models fitted to a representative cylinder, flow area to enable flow, tumble and swirl to be measured.
- Reciprocating Rigs – Laboratory testing of designs on single cylinder engines. These may be motored or fired depending on testing complexity. Access for diagnostic equipment is generally improved over full engine testing.
- Optical diagnostic techniques – Primarily the use of lasers and flow imaging systems to acquire data on the flow structure within models tested on either of the above testing rigs.
- Computational Fluid Dynamics (CFD) – the application of three dimensional mathematical models to the behaviour of fluid flows. Used to predict the efficiency and flow characteristics of inlet port and cylinder combinations.
- Full Engine Testing – Most valuable since this combines all the environmental factors which affect the system performance. This is the ultimate validation of any proposed design, but the other approaches benefit greatly in cost, data quality and flexibility.

A coherent strategy is clearly vital to balance the benefits of each development strategy against one another and maximise product value.

5. The Lotus Vortex Tumble Control System (V.T.C.S.) Inlet Port Concept

Consumer and legislative expectations for performance, economy and emissions of passenger car engines continue to rise. Unfortunately, the current base engine architecture of four-valve per cylinder, central spark plug, tumble augmented combustion system with fixed valve timing is beginning to approach the limit of cost effective expected return on investment.

In order to enhance the abilities of the powertrain product further a whole host of technologies are being tested and in high specification / low volume vehicles, now used in an attempt identify the most effective systems at minimum cost. Most of these systems work to augment or control the flow performance of the engine in some manner.

Figure 11 presents the typical targets for next generation passenger car engines alongside a contemporary engine. The greatest demands are for reduced emissions, improved efficiency in every area, increased specific power in order to downsize and reduce friction, and increased levels of low speed torque which allow the vehicle to run with longer overall gear ratios (to improve fuel economy). In order to meet these targets, it is clear from the analysis presented in the previous section that both the flow performance and the combustion system knock limit will need to improve. This is an extremely difficult target to meet given the highly developed standard of current passenger car S.I. engines using optimised tumble strategies. To develop a system to meet these challenges is a major focus of engine design groups around the world.

In this section, Lotus present a novel new concept for inlet port design which offers the potential to augment engine tumble and flow characteristics without the need for costly port deactivation systems or complex engine re-packaging. A port concept originated at Lotus has been developed, laboratory tested both via steady flow testing and CFD and is now in the process of patent application. Most importantly, the benefits of this concept are based around a manufacture feasible port feature applied as part of the geometry. This potentially negates the need for flow control valves, actuation systems and additional engine control strategies.

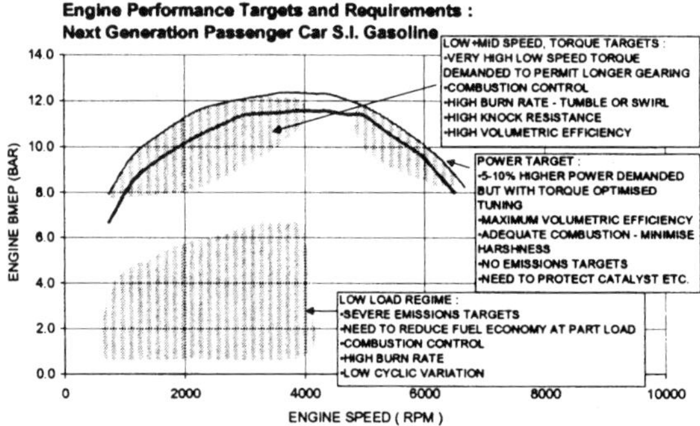

Fig.11. Targets for next generation passenger car SI gasoline engines – increased power and torque and lower emissions

5.1 Lotus Vortex Tumble Control System – Inlet Port Geometry
Measurements conducted at Lotus indicate that the flow and tumble performance of contemporary tumble ports of passenger car and high performance S.I. engines is closely linked to the flow separation mechanism on the floor of the port throat (Figure 12).

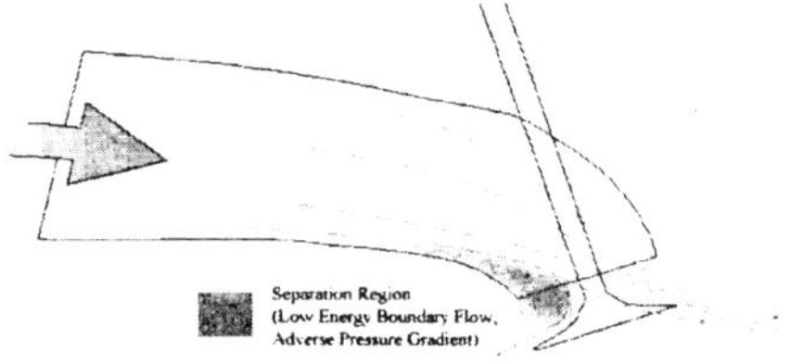

Figure 12: Tumble generation mechanism – Port
flow separation close to port/valve throat.

Fig. 12: Tumble generation mechanism – Port flow separation close to port/valve throat

A typical port geometry from a contemporary four-valve per cylinder engine generates initially a low level of tumble at low flow rates. As flow rate increases, the tumble increases markedly (Influence of increasing Reynolds number) and the flow coefficient decays. In the high speed, full load regime ports developed to deliver sufficient tumble in the low speed, low load regime will deliver substantially higher than necessary turbulence augmentation and lower flow coefficient (just when flow is required and turbulence levels are adequate through the inducted charge's momentum). The potential for excessive tumble is sometimes an important issue in high speed engines where towards the maximum engine speed, combustion may become highly unstable and is perceived as poor from a stability and harshness point of view. The loss of flow performance will reduce the engine's ability to generate power. Conventional strategies to reduce this effect will generally smooth the geometry around the separation region or raise the port approach angle, improving power and reducing harshness for a reduction in combustion quality and knock limit. The concept described in this section is intended to provide a new mechanism for flow and tumble control.

An example of the Lotus VTCS concept as applied to the inlet ports of a four valve per cylinder passenger car S.I. engine is illustrated in Figure 13.

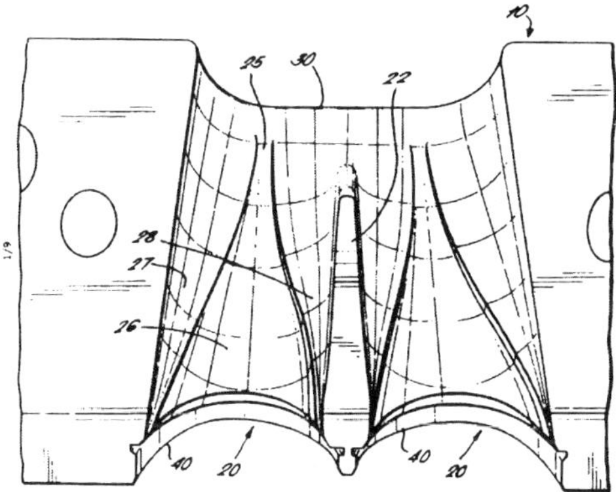

Fig.13. Vortex Tumble Control Control System – Inlet Port Floor Geometry (numbers referenced in text)

The design change is predominantly a diverging central duct form along the port floor (26), flanked by raised 'foil' portions (27/28) designed to interact with the boundary layers which exist close to the port surfaces. These layers contain the sluggish boundary flow which is responsible for flow phenomena such as flow separation effects. The depth of the duct on each port branch varies between 1 and 3mm, a small fraction of the port diameter and hence has little impact on the flow performance in terms of significantly diverting or deflecting the inviscid flow within the port.

It is the interaction of the duct edges with the inlet airflow during induction, and their varying impact on flow behaviour across the engine speed and load range, which is designed to augment and provide greater control over the engine performance and combustion.

This is achieved by augmentation of the port flow by controlled counter rotating vortex sheets as the flow approaches each valve (Figure 14).

These vortices (29) are optimised to entrain high energy flow from the free-stream into the boundary layer flows. At low flow rates the impact of these devices was found to be low. At higher flow rates, when flow separation becomes well developed, the vorticity is used to augment the flow structure by re-establishing flow control, delaying the on-set of separation in the throat area, and thus reducing tumble levels and enhancing flow performance. To validate the concept, analysis using steady flow rigs and computational fluid dynamics was conducted.

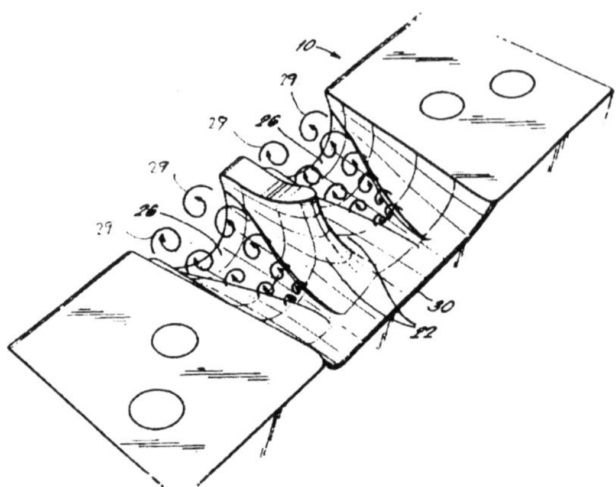

Fig.14. Representation of flow field generated by VTCS in port throat region

5.2 Steady Flow Measurements
Models of the modified and standard port geometry were tested on a Superflow 600 steady flow laboratory flow rig, across a range of air pressure differentials whilst measuring the flow and tumble parameters for the full range of valve lift. Investigation of the in-cylinder flow via simple optical techniques confirmed the presence of the vortex structures.

The results are presented in Figures 15 and 16 below for flow coefficient and tumble levels respectively. The edge radius for the VTCS port used in these tests was 0.2 mm.

Figs. 15 and 16 indicate that while flow performance and tumble levels for both ports were very similar at low to mid flow rates; when the flow rate was increased the modified port achieved higher flow coefficients and a reduced rate of tumble. The mean flow coefficient improved by 5%, 16% at maximum lift and, while tumble at low flows was very similar, the tumble was reduced by 7% at maximum tested flow rate. This is a significant improvement in flow coefficient considering the well developed nature of the base geometry. Fig. 17 shows that the flow coefficient for base port reduced by 3% over the tested flow range. By using sharp radii on the edges of the VTCS profile, the flow coefficient was recorded to increase between 2 and 4%, from low flow rate to maximum flow rate performance (Fig. 17). This improvement was matched by a reduction in Tumble for all design variants tested with the exception of those with foil edge radii in excess of 2mm as shown in Fig. 18. It was also noted that the trend was for tumble to continue to diminish as flow increased. The rate of drop of was closely related to the radii employed at the edges of the profile.

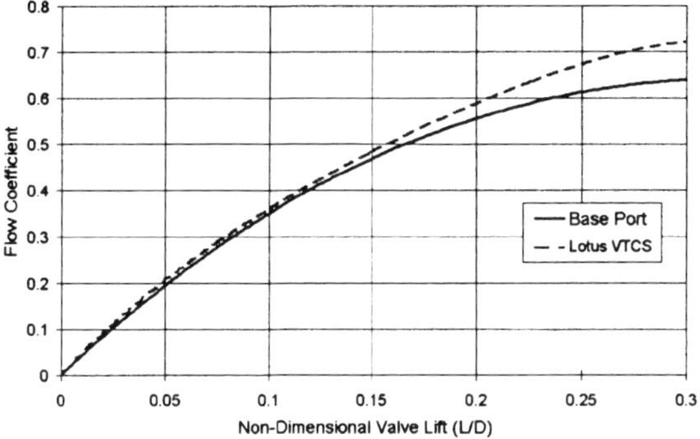

Fig.15. Variation of flow coefficient for standard and Lotus VTCS port with valve lift

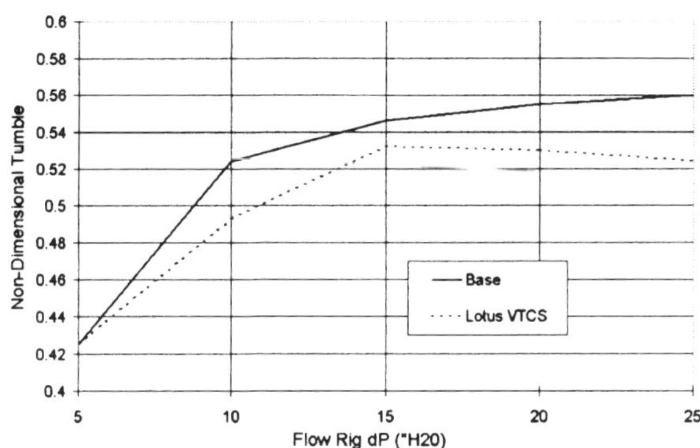

Fig.16. Variation of non-dimensional tumble with pressure drop across port

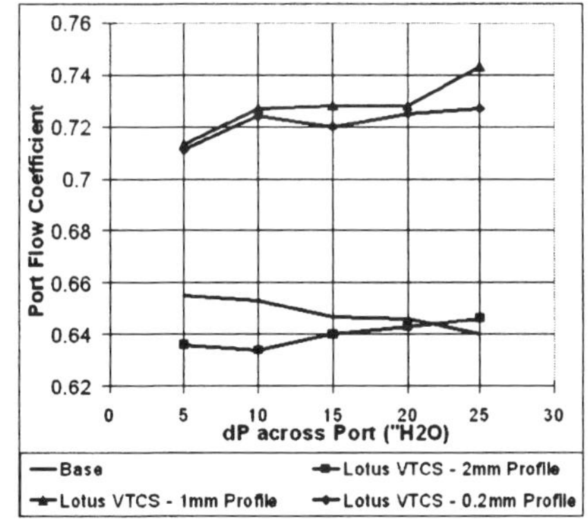

Fig.17. Variation of flow coefficient for standard and Lotus VTCS port with pressure drop across the port

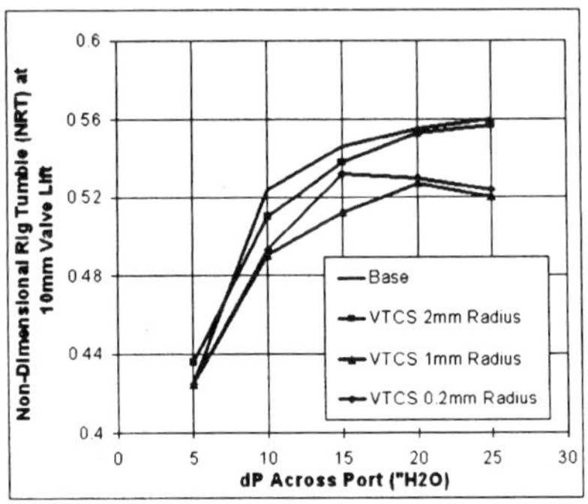

Fig.18. Variation of non-dimensional tumble with pressure drop across port for different edge radii on VTCS profile

5.3 CFD Analysis

CFD analysis was conducted by Lotus Powertrain CAE using Star-CD. Simulation compared the predicted steady flow performance of the base port model with a modified version adopting the Lotus VTCS concept. Modelling employed ICEM-Tetra to generate a tetrahedral mesh of $>2 \times 10^5$ cells (Figure 19). Solution was conducted with the Mars solver.

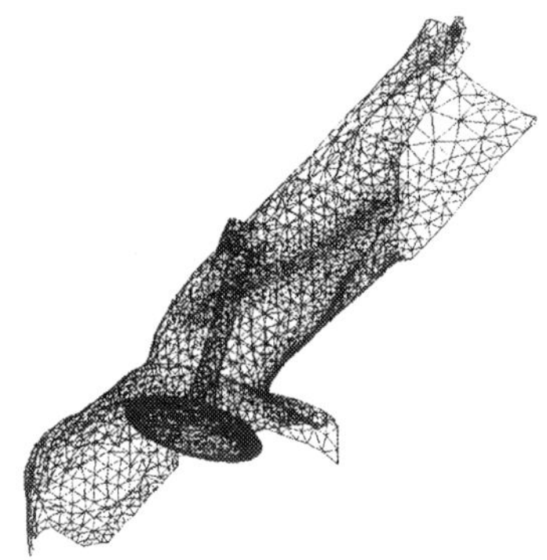

Fig.19. Tetrahedral mesh for Star-CD CFD simulation – Inlet Port Half Model

A steady flow analysis was carried out for both cases and the results compared. Simulation achieved excellent correlation with the steady flow laboratory testing, predicting an improvement in flow coefficient of +8% and demonstrating the presence of in-port vortex flows (Figure 20).

In addition to the presence of the vortex sheets, the CFD analysis also highlighted the thinning of the boundary layer in the channel, the entrainment of flow into the channel region and delayed separation of the flow in the throat region. The very low pressure and high local turbulence rates would likely have a significant potential to enhance fuel preparation. This is effect may be highly beneficial since the uncontrolled fuel droplets associated with MPFI engines tend to disperse from the injection spray toward the port floor. When entrained into the vortex regions the low pressure core floors should have a positive impact on the atomisation of these droplets.

These results and their close agreement with both the conceptual ideas and results of physical testing have been very encouraging. Simplified transient CFD analysis has also been conducted. This study was used to evaluate the validity of induced vorticity in a simulated flow similar to that of a running engine. The analysis confirms that the action of these foils and vortex generation is well within the time scales of the engine induction event.

Lotus is currently planning to conduct further analysis and physical testing in order to explore the value of this technology and an application for patent has been made.

The value of the VTCS concept for future engines would be in enabling engines to extend the trade-off limits between tumble and flow. It is possible that the operation of VTCS in the very high flow rate environment at maximum engine speeds will provide significant gains in engine performance such that the tumble features of the engine can be radically increased for improved low speed, low load emissions and efficiency, without generation of excessive tumble at maximum power. This would be a valuable, uncomplicated and low cost contribution to the improvements in performance, economy, emissions and refinement, which engines of the future will no doubt achieve.

5. Conclusions

The use of multiple poppet valves, and the coupled development of intake flow and combustion technology, has enabled a variety of strategies to evolve which provide a route for meeting the performance, fuel economy, and emissions targets of modern engines.

In gasoline engines the use of four-valve per cylinder configurations with pent-roof combustion chambers offers a compromise between obtaining a large gas flow area across the speed range and achieving high turbulence levels, and thus flame speeds, at low engine speeds and loads. The Lotus VTCS is a low cost approach which mitigates the level of compromise in flow area for a given tumble ratio.

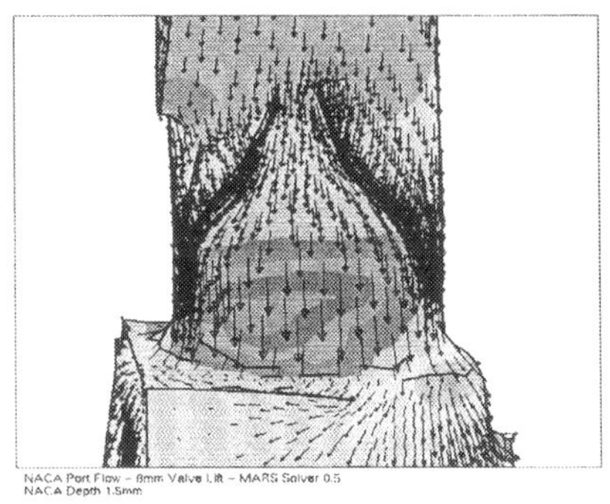

Fig.20. Features in flow predicted using CFD

References

1. **Monaghan, M.L.**, In: *Internal combustion engines.* Ed. Arcoumanis, C., Combustion Treatise, Academic Press, 1988.
2. **Ricardo, H.R.**, *Engines of high output: thermodynamic considerations.* Second Impression. MacDonald and Evans, London, 1929.
3. **Borgeson, G.**, *The classic twin-cam engine.*
4. **Lovell, B.**, *Car Design and Technology*, pp. 70-73, Brackland Publishing, March 1992.
5. **Rudd, A.C.**, *How many valves?* I. Mech. E. International Seminar: Powertrain - Engine Concepts. N.E.C. Birmingham, 1985.
6. **Lichty, L.**, *Internal combustion engines.* Fifth Edition. McGraw-Hill Book Co., New York, 1939.
7. **Ricardo, H.R.**, *Piston aero-engines.* Proc. Instn. Mech. Engrs., 10th June, 1947. (In *An engineering archive.* Eds. Winterbone, D., and Moore, K., Mechanical Engineering Publications, 1997.
8. **Willoughby, V.**, *Classic Motorcycles.* Hamlyn, 1976.
9. **Yagi, S., Ishizuya, A., and Fujii, I.**, *Research and development of high-speed, high-performance, small displacement Honda engines.* SAE paper 700122.
10. **Amann, C.A.**, Discussion - *Flow area of multiple poppet valves.* Proc. Instn. Mech. Engrs., 211, Part D, 407-408.
11. **Treue, W., and Zima, S.**, *Hochleistungsmotoren - Karl Maybach und sein Werk*, VDI-Verlag, Dusseldorf, 1992.
12. **Ohrnberger, G., and Mann, M.**, *The Audi 5-valve cylinder head concept.* SAE paper no. 945004, 1994.
13. **Robson, G.**, *Cosworth - The search for power.* Patrick Stephens Ltd., 1991.
14. **Fukui, T., Satoh, I., and Nishiyama, R.**, Realization of high-performance motorcycle engine 8-valve system racing engine. J. SAE Japan, vol. 39, no. 9, 1985.

15. **Kanazawa, S., Akagi, M., Fukumachi, M., and Suzuki, M.,** Development of elliptical piston engine for motorcycle. SAE paper no. 930224, 1993.

16. **Livengood, J.C., and Stanitz, J.D.,** *The effect of inle-valve design, size, and lift on the air capacity and output of a four-stroke engine.* NACA TN no. 915, 1943.

17. **Fukutani, I., and Watanabe, E.,** *An analysis of the volumetric efficiency characteristics of 4-stroke cycle engines using mean inlet Mach number Mim.* SAE paper 790484, 1979.

18. **Taylor, C.F.,** *The internal combustion engine. Vol. 1.* Second. Edition, MIT Press, Cambridge, Mass., 1966.

19. **Pashley, N.,** *Limitations on volumetric efficiency of internal combustion engines.* Final Year Project Dissertation, Department of Mechanical Engineering, UMIST, Manchester, 1993.

20. **Woods, W.A., and Brown, P.G.,** *Flow area of multiple poppet valves.* Proc. Instn. Mech. Engrs., Part D, vol. 210, 347-352, 1996.

21. **Aoi, K., Nomura, K., and Matsuzaka, H.** *Optimization of multi-valve, four cycle engine design - the benefit of five-valve technology.* SAE paper no. 860032, 1986.

22. **Downing, I.C.,** *Design and development of the Tickford five-valve cylinder head.* Autosports International Congress, NEC, Birmingham, Jan. 3rd-6th, 1991.

23. **Sykes, R.,** Development of a 5 valve per cylinder variant of the Ford Duratech 2.5 litre V6 engine, Autotech 95 paper no. C498/22/174.

24. **Joyce, M.J., Carling, J., Clough, M., and Corkhill, W.J.,** *A new Jaguar engine for the 21st century.* I. Mech. E. paper no. C517/003/96, International Seminar on Application of Powertrain and Fuel Technologies to Meet Emissions Standards, 24th-26th June, 1996.

25. **Crouch, A.R., Gilbert, I.P., and Norris, R.D.,** *Gasoline engine efficiency - design solutions.* I.Mech.E paper no. C389/188, 1992.

26. **Rothrock, A.M., and Spencer, R.C.,** *The influence of directed air flow on combustion in a spark-ignition engine.* NACA Report no. 657, 1939.

27. **Lee, D.W.,** *A study of air flow in an engine cylinder.* NACA Report no. 653, 1939.

28. **Haworth, D.C., El Tahry, S.H., Huebler, M.S., and Chang, S.,** *Multidimensional port-and-cylinder flow calculations for two- and four-valve-per-cylinder engines: influence of intake valve configuration on flow structure.* SAE paper no. 900257, 1990.

29. **Floch, A., van Frank, J., and Ahmed, A.,** *Comparison of the effects of intake-generated swirl and tumble on turbulence characteristics in a 4-valve engine.* SAE paper no. 952457, 1995.

30. **Neuber, H.-J., Spiegel, L., and Ganser, J.,** *Particle tracking velocimetry - a powerful tool to shape the in-cylinder flow of modern multi-valve engine concepts.* SAE paper no. 950102, 1995.

31. **Kang, K.-Y., Oh, S.M., Lee, J.-W., Jeong, D.-S., Lee, K.-H., and Bae, C.-S.,** *The effect of intake port design in lean burn SI engines.* I.Mech. E. paper no. S433/009/96, 1996.

32. **Borman, G.L., and Ragland, K.W.,** *Combustion engineering.* WCB/McGraw-Hill, Boston, 1998.

33. **Tindal, M.J., Brown, P.G., and Kyriakides, S.C.,** *An investigation of swirl and turbulence in the cylinders of direct injection diesel engines.* Conf. On Diesel Engines for Passenger Cars and Light Duty Vehicles, I.Mech.E., London, 1982.

34. **Horie, K. and Nishizawa, K.,** *Development of a high fuel economy and high performance four-valve lean burn engine.* I.Mech.E paper no. C448/014, 1992.

35. **Wilson, N.D., Watkins, A.J., and Dopson, C.,** *Asymmetric valve strategies and their effect on combustion.* SAE paper no. 930821, 1993.

Multi-valve cylinder heads for uniflow scavenged two-stroke engines and their effect on naturally aspirated and turbocharged performance

F J WALLACE
Department of Mechanical Engineering, University of Bath, UK

1) <u>Introduction</u>

The effect on performance of replacing the traditional 2 valve cylinder head in 4 stroke engines by multiple inlet and exhaust valves, generally in the form of 2 valves of each type, has been described in the literature and is now very well understood. (1, 2) The increase in available flow area has beneficial effects on the gas flow through the engine and hence on both output and efficiency. Furthermore multiple inlet vales in conjunction with directed inlet ports permit close control of inlet induced swirl and hence of in-cylinder air motion with further beneficial effects on fuel-air mixing.

Potentially the benefits obtainable from multiple valves in exhaust valve-in-head, Uniflow scavenged 2 stroke engines, both of the gasoline and diesel variety are even greater in view of the greater sensitivity of air throughput to the relevant port and valve arrangements.

This paper investigates the effect of 4 versus 2 exhaust valves on the performance of a hypothetical 3 cylinder 2 stroke direct injection diesel engine of 82.5 mm bore and 82 mm stroke (3), having 6 directed inlet ports of 22.9 mm width and 14.1 mm height, and either 2 exhaust valves of 36.4 mm diameter or 4 valves of 30.2 mm diameter, in line with the geometric analysis of alternative valve arrangements by Woods (4).

The performance assessment refers to two alternative builds of the engine, viz.

a) with the engine equipped with a geared high efficiency turbocharger, the gearing allowing any power deficit between turbine and compressor to be made good by abstracting some power from the crankshaft, and

b) with the engine operating essentially under naturally aspirated conditions, the air being supplied by a mechanically driven blower to overcome the inevitable pressure drop across the engine. (See fig 1 for layout)

The assessment is based on a cycle simulation program developed at Bath University (5), containing 2 stroke scavenging model which divides the process into two phases, an initial perfect displacement phase in which incoming air physically displaces the products of combustion in the cylinder, followed by a perfect mixing phase in which every increment of incoming air mixes simultaneously with the residual cylinder charge, the proportions of the total air supplied taking part in each phase being defined by a perfect displacement fraction x and a perfect mixing fraction $(1 - x)$. There is also provision for a short circuiting fraction y, but this is considered negligible in Uniflow scavenged engines (6).

The process is illustrated diagrammatically in fig 2, showing efficiency defined as:

$$\eta_{ch} = \frac{mass\ of\ delivered\ air\ retained}{displacement\ volume\ x\ inlet\ manifold\ density}$$

as a function of delivery ration, defined as

$$DR = \frac{mass\ of\ delivered\ air}{DISPLACEMENT\ volume\ x\ inlet\ manifold\ density}$$

For various values of the displacement, mixing and short-circuiting fractions.

2) <u>Performance Assessment</u>

2.1)   General

For the purposes of this paper the assessment applies to the limiting torque curve, over a speed range from 1000 rev/min to 3500 rev/min, since this makes maximum demands on engine performance.

Figures 3 to 7 illustrate the most important performance and scavenge parameters for the two engine builds, i.e. turbocharged (fig 1a) or naturally aspirated, with blower assist (fig 1b), both in 2 or 4 exhaust valve form.

The figures will be discussed in the following order:

Fig 3: Torque (Nm)
Fig 4: Nett power and power abstracted from crankshaft to assist the turbocharger or
      drive the blower (kW)
Fig 5: Brake thermal efficiency (%)
Fig 6: Delivery ratio
Fig 7: Charging efficiency

In all cases a schedule of trapped air-fuel ratio as a function of engine speed has been assumed, starting at 27 : 1 for rated conditions at 3500 rev/min and reducing to 20 : 1 at the peak torque point at between 2000 and 2500 rev/min, below which it remains constant. In the case of the turbocharged engine, the gear ratio between crankshaft and turbocharger is adjusted to obtain a schedule of boost (i.e. inlet manifold pressure) designed to give an optimal torque curve, common to both the 2 and 4 exhaust valve builds, while in the case of the naturally aspirated (mechanically blown) engine, the gear ratio which is fixed for both the 2 and 4 valve builds, has been adjusted to give similar boost schedules for both builds. (1 : 1 for the 2 valve build and 1.115 : 1 for the 4 valve build).

2.2)   Discussion of performance and scavenge characteristics

Fig 3: <u>Limiting torque</u>

The turbocharged engine with a maximum boost of 1.98 bar abs at 3500 rev/min as against only 1.22 bar for the naturally aspirated engine, operates at much higher torque levels throughout, with peak torque of 264 Nm (4 valve) and 244 Nm (2 valve) at 2000 rev/min, while the naturally aspirated engine develops only 153 Nm (4 valve) and 139 Nm (2 valve) at 2500 rev/min.

The lower torque levels of the 2 valve build in each of the above cases are accounted for by the lower airflow, and hence delivery ratio (see fig 6) at any given boost level due to the higher flow resistance, with its resultant reduction in charging efficiency (fig 7, also fig 2) and the associated lower mass of trapped air.

Fig 4: <u>Nett power and power abstracted from crankshaft for turbocharger power assist or to drive blower</u>

As with torque, the turbocharged engine achieves much higher power levels than the naturally aspirated engine, viz. 69.57 kW (4 valve) and 62.28 kW (2 valve) at 3500 rev/min, as against 43.62 kW (4 valve) and 39.22 kW (2 valve) for the naturally aspirated engine at the same speed.

The discrepancy between the 4 valve and 2 valve builds remains significant over the whole speed range both for the turbocharged and naturally aspirated builds, although absolute differences reduce with decreasing speed and power level. As with torque, the advantages of the 4 valve build are clearly seen.

The lower part of fig 4 shows the level of power abstracted from the crankshaft for turbocharger assist in the case of the turbocharged engine - the turbocharger in high speed 2 stroke engines generally cannot be fully self sustaining - or the power abstracted to drive the blower in the case of the naturally aspirated engine.

In the case of the turbocharged engine the level of power assist is very small for the 4 valve build due to its low pressure losses, with a minimum of only 0.13 kW at 2500 rev/min when the power circulating in the turbocharger is 7.76 kW. The turbocharger is thus virtually self sustaining. For the 2 valve build, on the other hand, the minimum level of power assist is 1.17 kW at the same engine speed. As engine speed is reduced below 2500 rev/min, the level of power assist rises considerably, to 2.02 kW (4 valve) and 2.63 kW (2 valve) both valves now being a significant proportion of the power circulating in the turbocharger, of approximately 5 kW, at 1000 rev/min engine speed and boost of 1.72 bar (abs).

In the case of the naturally aspirated engine, blower power demand rises continuously with engine speed, with increasing air mass flow leading to increasing pressure drop across the engine. Maximum power demand at 3500 rev/min is 2.96 kW (4 valve) and 2.64 kW (2 valve) at a common boost level of 1.22 bar abs, but with the mass flow for the 4 valve engine being somewhat higher due to the larger effective flow areas.

Fig 5: <u>Brake thermal efficiency</u>

The turbocharged engine achieves it best efficiency of 40.26% (4 valve) and 38.82% (2 valve) at 3500 rev/min, i.e. rated conditions. Efficiency is maintained to within 1 percentage point down to 2500 rev/min, but then drops gradually to 30.10% (4 valve) and 28.89% (2 valve) at 1000 rev/min, when turbocharger power assist is a maximum. The lower efficiency of the 2 valve build over the whole speed range is due partly to its lower delivery ratios (fig 6) and hence charging efficiency (fig 7) and partly due to the higher level of power assist resulting from higher pressure drops across the engine, and thus lower available turbine power.

The naturally aspirated engine shows a significantly lower level of efficiency for both 4 valve and 2 valve builds compared with the turbocharged engine, due to the higher proportion of engine power required to drive the blower. Best efficiency now occurs at 2500 rev/min, being 36.47% (4 valve) and 36.23% (2 valve), similar small differences applying over the whole speed range. This is due to the fact that the higher engine efficiency (as opposed to the overall efficiency including blower power) resulting from the superior scavenge characteristics of the 4 valve engine, is offset by the higher required blower power due to the higher air mass flow, as already discussed in connection with fig 4.

Fig 6: <u>Delivery ratio</u>

Delivery ratio, defined as:

$$DR = \frac{mass\ of\ delivered\ air}{DISPLACEMENT\ volume\ x\ inlet\ manifold\ density}$$

is a measure of the air throughput through the engine which has a vital bearing on the scavenge quality, expressed in this paper by the charging efficiency η_{ch} (see fig 7). Fig 6 shows opposed trends of delivery ratio for the turbocharged and naturally aspirated builds, due to the quite different characteristics of the centrifugal compressor of the former, and the rotary positive displacement blower of the latter. In a plot of delivered pressure vs mass flow, the constant speed lines of the centrifugal compressor ran almost horizontally over a large part of the available flow range, while those of the positive displacement blower ran almost vertically, the deviation from verticality being due to tip leakage.

The superiority of the 4 valve build is clearly apparent both for the turbocharged and naturally aspirated builds. For the former, delivery ratio decreases from 2.071 (4 valve) or 1.746 (2 valve) at 1000 rev/min to 0.814 (4 valve) or 0.801 (2 valve) at 3500 rev/min.

For the naturally aspirated engine, on the other hand, delivery ratio increases from 1.028 (4 valve) or 0.913 (2 valve) at 1000 rev/min to 1.156 (4 valve) or 1.024 (2 valve) at 3500 rev/min. The fact that delivery ratio thus remains close to unity over the whole speed range is the direct result of the fact that with positive displacement blowers, mass flow is almost directly proportional to speed which, in turn is directly related to engine speed, due to the fixed gear ratio employed [1.115 (4 valve) and 1.0 (2 valve)].

The higher delivery ratios of the 4 valve build both in turbocharged and naturally aspirated form lead directly to higher levels of charging efficiency (fig 7), i.e. higher levels of retained fresh air charge which, in turn, lead to higher levels of power and efficiency.

Fig 7: <u>Charging efficiency</u>

Charging efficiency has already been defined as:

$$\eta_{ch} = \frac{mass\ of\ delivered\ air\ retained}{displacement\ volume \times inlet\ manifold\ density}$$

In the case of a simplified isothermal, constant volume, incompressible scavenge process, it is possible to derive an analytical relationship between charging efficiency and delivery ratio in terms of assigned values of the displacement, perfect mixing and short circuiting fractions, x, z and y respectively as follows:

$$\eta_{ch} = 1 - (1 - x)\,e^{(1 - y)(x - DR)}$$

where x = displacement fraction
             y = short circuiting fraction
             DR = delivery ratio

Numerical values are shown, as a function of η_{ch} vs DR, for different values of the fractions x and y, (in fig 2) which shows very clearly the beneficial effects of large values of delivery ratio and displacement fraction and the adverse effects of short circuiting. The above treatment has been adapted to the much more complex situation in engines where pressure, temperature and composition all vary. Nevertheless it has been possible in the simulation program to retain the concept of displacement, mixing and short circuiting fractions. For the present investigation values of 0.6 and 0.5 based on the predicted quality of scavenging have been selected for the displacement fraction for the case of the 4 valve and 2 valve engine respectively, with the short circuiting fraction set to zero, as is appropriate in the case of the Uniflow scavenged engines. While values of charging efficiency predicted by this more extensive treatment differ somewhat from those derived from the simple analytical treatment, the trends are very similar.

Referring to fig 7, charging efficiency for the turbocharged engine <u>reduces</u> from 0.8391 (4 valve) or 0.7904 (2 valve) at 1000 rev/min to 0.6413 (4 valve) or 0.5852 (2 valve) at 3500 rev/min, while for the naturally aspirated engine they <u>rise</u> from 0.7105 (4 valve) or 0.6586 (2 valve) at 1000 rev/min to 0.7284 (4 valve) or 0.6553 (2 valve) at 3500 rev/min. The opposing trends are due to the fact that, as already discussed, delivery ratio (fig 6) <u>reduces</u> with increasing speed with the turbocharged engine, while it <u>increases</u> with the naturally aspirated engine.

In relation to engine performance, charging efficiency also determines the level of charge purity, i.e. the mass fraction of trapped pure air in the cylinder, and thus of the mass fuel burnt at any given trapped air/fuel ratio. The higher levels of charging efficiency of the 4 valve engine thus imply higher levels of engine power and efficiency, as previously discussed in connection with figs 4 and 5.

3) <u>Summary and Conclusions</u>

The performance parameters described viz.

a) limiting torque (fig 3)
b) power (fig 4)
c) system brake thermal efficiency (fig 5)

all show significant improvements, of up to 10%, both for the turbocharged and naturally aspirated builds, resulting from the use of cylinder heads with 4 rather than 2 exhaust valves. These gains are directly attributable to the improvement, at any given level of engine speed, boost and trapped air/fuel ratio, of delivery ratio and the consequential increase in charging efficiency (figs 6 and 7). It is suggested that the potential gains from the use of multi valve heads are even greater in 2 stroke engines than in the well known case of 4 stroke engines.

4) <u>Appendix</u>

<u>Charging efficiency</u>

The derivation of the analytical expression for charging efficiency for the simplified case of isothermal, constant volume, incompressible flow is given in detail in reference 6, leading to

$$\eta_{ch} = 1 - (1-x)\,e^{(1-y)(x-DR)} \qquad (1)$$

and may be regarded as an extension of the classical Hopkinson expression for an equivalent perfect mixing process, viz. (with $x = 0$, $y = 0$)

$$\eta_{ch} = 1 - e^{-DR}$$

The incorporation of a similar approach in the cycle simulation program, with full allowance for continuous changes, during the scavenging process, of pressure, temperature and composition, has made possible the detailed prediction of changes in performance due to the marked effect of alternative exhaust valve arrangements on delivery ration and charging efficiency.

References

1. Lovell, B.: "Four valve cylinder heads", Design and Technology, March 1992, pp. 70-73.

2. Amann, C.A.: Contribution to paper by W.A. Woods (ref 4), IMechE 1997.

3. Cox, A., Wallace, F.J., Bird, G. and Horrocks, R.: "Initial Development of an Experimental Uniflow Two Stroke DI Diesel Engine", IMechE, Autotech 95, Paper C 498/2/039.

4. Woods, W.A.: "Flow Areas of Multiple Poppet Valves", Technical Note, Institution of Mechanical Engineers, vo. 210, 1997, pp 347-351.

5. Wallace, F.J.: Presentation of Simulation Package ODES at Universities Internal Combustion Group (UNICEG) meeting, University College, London, December 1993.

6. Dang, D., Wallace, F.J.: "Some Single Zone Scavenging Models for Two Stroke Engines", Int.J.Mech.Sci, Vol 34, No 8, pp 595-604, 1992.

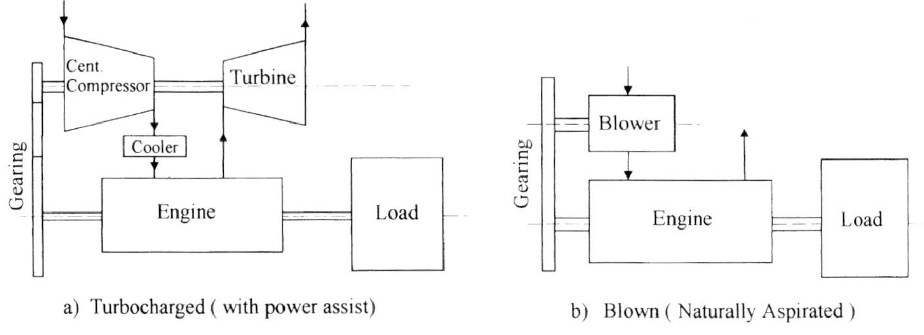

a) Turbocharged (with power assist)   b) Blown (Naturally Aspirated)

FIG 1   Engine Builds

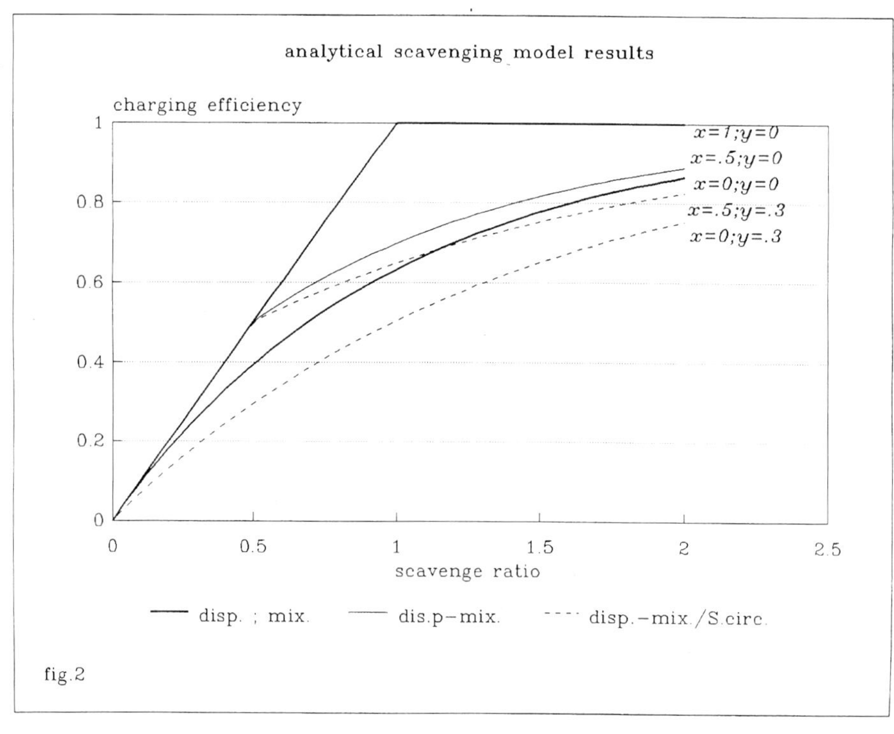

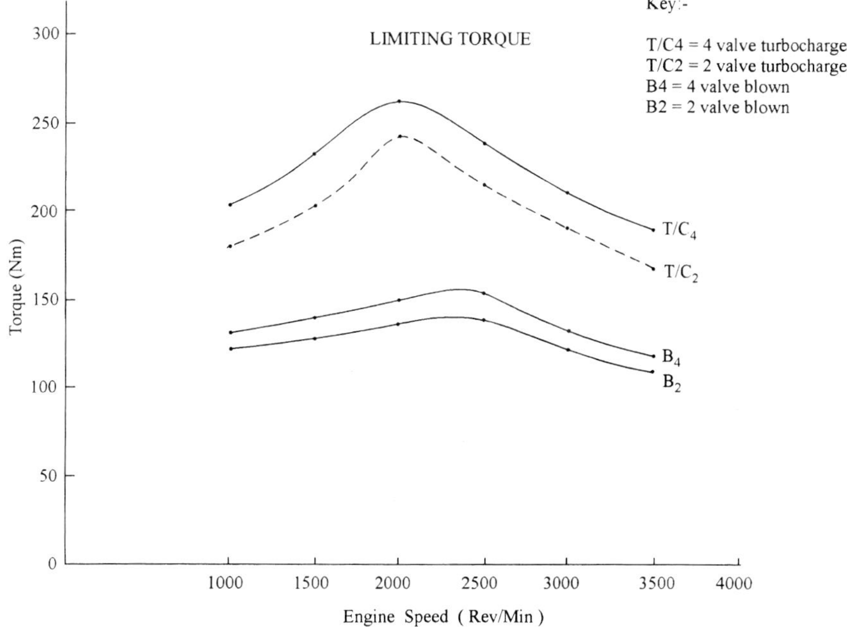

FIG. 3

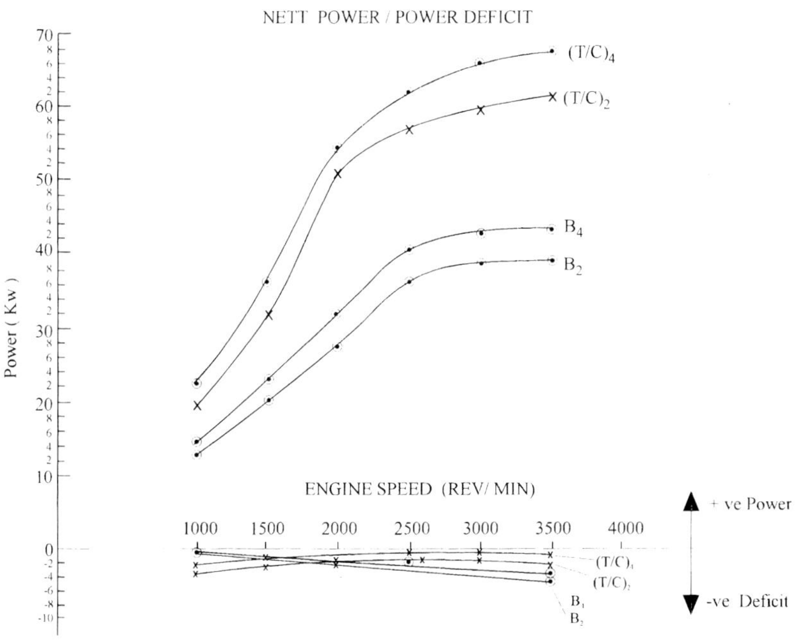

FIG 4

SYSTEM EFFICIENCY
Efficiency (%)
40
35
30
25
20
(T/C)4
(T/C)2
B4
B2
0   1000   1500   2000   2500   3000   3500   4000
Engine Speed (Rev/Min)

FIG 5

DELIVERY RATIO
Delivery Ratio
T/C4
T/C2
B4
B2
0   1000   1500   2000   2500   3000   3500   4000
Engine Speed (Rev/Min)

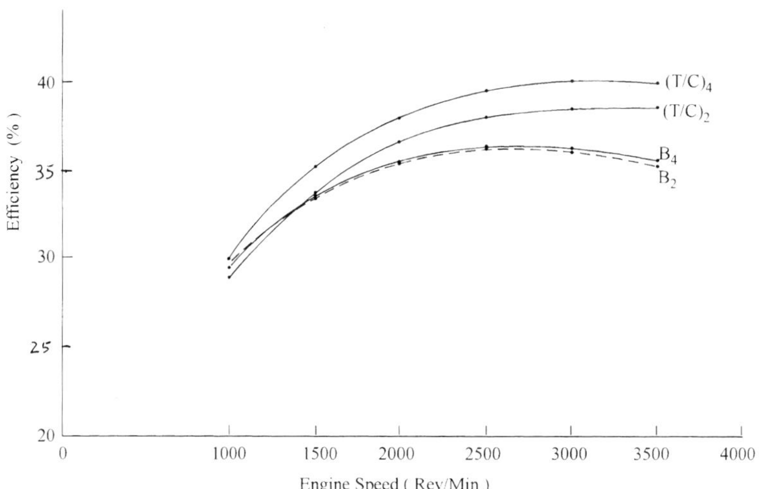

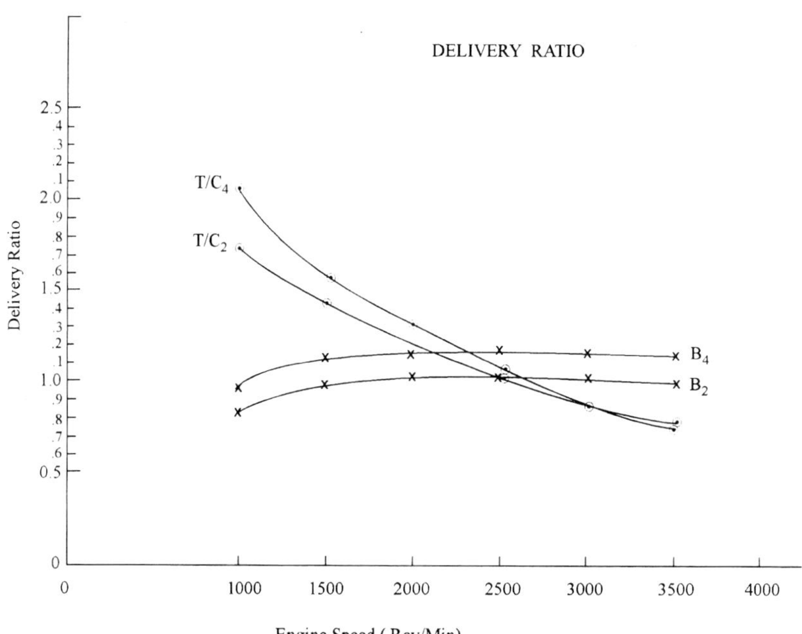

FIG 6

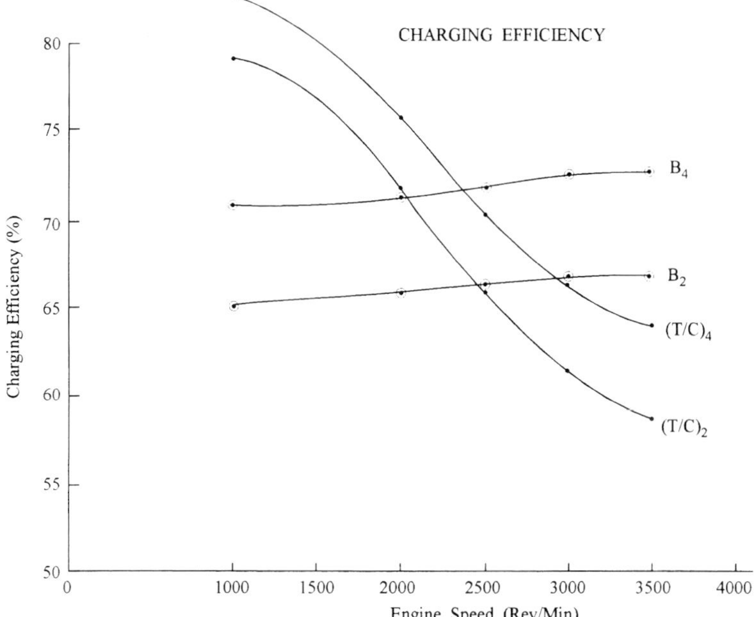

FIG 7

Ranking the flow capacity for the design of multi-valve schemes

W A WOODS
Queen Mary Westfield College, University of London,UK
A M WOODS
School of Mathematics, University of Bristol, UK

Synopsis

The primary purpose for poppet valves in reciprocating engines is to facilitate the gas exchange process. The basic factors affecting the flow are considered. The two principle parameters are the effective flow area of the valve and the upstream pressure. As the inlet pressure is fixed on naturally aspirated engines, the inlet valves are designed larger than those of the exhaust valves. Although multi-valve schemes have been used for competition purposes and their advantages have been known for many years, it is only in recent times that the market penetration of engines with four valves per cylinder has overtaken that of the conventional two. The objectives of this paper are to present and discuss major extensions to the simple method of ranking the flow capacity of multi-valve schemes [6]. The paper presents a much more relevant treatment than loc cit by including work on pent roof cylinder heads with off-set designs between inlet and exhaust sections. Configurations with two, four and six valves are analysed and a technique is devised for considering schemes with odd numbers of valves, typically five as used on some recent high performance engines. The results of the investigation are first presented separately for the inlet and the exhaust values, in a logical and systematic manner, but to assist in making trade-off judgements the results are presented on one diagram for both the inlet and exhaust valves. The superior cases emerge as those with three inlet valves and three exhaust valves and those with three inlet valves and two exhaust valves. However, before more definite conclusions can be drawn, this work needs to be backed up with engine simulation studies over a range of inlet to exhaust valve area ratios to find the optimum.

Notation

a half major axis of ellipse

b half minor axis of ellipse, radius of cylinder

c off-set measured along major axis of ellipse

d minimum diameter of port

l valve lift

n number of valves being considered

r radius of generic circle of valve

t clearance between valves or valve and cylinder wall

x, y cartesian coordinates

A area

P pressure

R ratio of cylinder

$\mathcal{R}$ gas constant

α angle of inlet valve projection surface shown on figure 4

β angle shown on figure 5, 6

γ angle shown on figure 6, ratio of specific heats

ϵ diameter of valve stem

θ angle of seat of valve

$\xi = c/a$

ϕ ellipse parameter angle shown on figure 5, 6.

Φ Non-dimensional area of one valve

ψ angle of exhaust valve projection surface shown on figure 4

$\Omega = 4n_j(r/b)^2$ generic area ratio of inlet $j = i$ or exhaust $j = e$ valves

Subscripts

d downstream

eff effective flow area of valve

e exhaust valve

i inlet valve

L pertaining to exhaust vale projection surface

O stagnation condition

1. Introduction

The primary purpose of poppet valves in engines is to facilitate the gas exchange processes. The flow rate through the valves depends mainly upon the effective flow area of the valves, the upstream stagnation pressure and the reciprocal of the upstream stagnation temperature. At higher pressure ratio across the valve, the flow is choked and this condition can occur on both inlet and exhaust valves.

The effective flow area and the upstream stagnation pressure are the parameters of major influence. In the case of the exhaust valves, the higher the cylinder pressure the greater the flow rate, hence there is a self-regulating effect, but in the case of the inlet valves on naturally aspirated engines, increasing the flow rate has to depend largely on increasing the effective flow area of the valves. Normally inlet valves have geometrical areas ranging from 36% to 53% larger than those of the exhaust valves [1].

A secondary purpose of inlet valves is to direct the inlet jet flows into the cylinder to produce large scale prescribed motion in order to enhance the subsequent combustion process. Examples are swirl and tumble [2]. In the case of exhaust valves, possible targets are to promote flow from regions of low unburnt hydrocarbons.

Multi-valve configurations lead to smaller valve diameters and therefore to smaller reciprocating masses. The mass of gemoetrically similar valves scales in proportion to d^3 but the flow area depends on d^2. An important consideration is that the smaller valves can not only operate at a higher engine speed but also at a higher lift to diameter ratio. With these features there are many possibilities for selective operation, such as that used on the Honda VTEC engine [3].

An early example of a multi-valve scheme is the four valve pent roof engine in the Peugeot racing cars of 1912 [2]. An interesting example of a successful pent roof design is the Mercedes Benz M196/110 engine with desmodromic valve gear of 1955. This example, taken from reference [4], is shown in figure 1. The large angles of inclination of the valves is a notable feature. In the United States, the market penetration of the four valve engine exceeded that of the two valve engine in 1994 and an example of a British Leyland four valve pent roof design taken from ref [5] is shown in figure 2.

The objectives of the present paper are to present and discuss an extension of the simple method of assessing the flow capacity of multi-valve schemes [6]. The new aspects addressed are the pent roof arrangements with different angle of valve plane inclinations and different off-set fractions between inlet and exhaust sections.

Initially the basis of the anlysis is reviewed and the non-dimensional effective area Φ is introduced and the generic area ratio Ω is defined. The off-set ratio ξ is introduced and interestingly, it is shown that both inlet and exhaust sections of the head have the same numerical value. The arrangements for one, two and three valves per section are considered and their combination to form various other multiple valve configurations is discussed.

Finally, to assist in trade-off judgements, the merits of various scehems are discussed with reference to an inlet valve area $\sim$ exhaust valve area diagram on which contours of inlet to exhaust valve area ratio are shown. Theoverall findings are discussed and superior results for the six and five valve arrangements are highlighted.

2. Basis of the Analysis

The effective flow area of a poppet valve A_{eff} is the area which, when used with the steady flow equations of the constant pressure model, gives the steady mass flow rate, as discussed in reference [7]. This area is a function of the generic area of the valve seat $\pi d^2/4$, the valve lift l and a number of other parameters including the pressure ratio across the valve, P_o/P_d, the ratio of the specific heats, γ, the gas constant $\mathcal{R}$, the valve seat angle, θ, the clearance between valves or valves and cylinder wall, t, and the diameter of the valve stem ϵ. The parameters of major concern in this work are the effective area ratio $\Phi = \frac{A_{eff}}{\pi d^2/4}$ and the valve lift ratio l/d. Typical results for this relationship are shown in figure 3. Although there are small differences between inlet and exhaust valves given in ref [7], these differences are neglected in the discussion which follows.

Following the definition given in ref [6], the generic area ratio Ω is given by the ratio of (the generic area of one valve times the number of valves being considered) and (the generic area of one valve of the two valve case for a flat head with equal diameter exhaust and inlet valves).

$$\Omega = 4n_j(r/R)^2$$

where $j = i$ or e.

The basic arrangement of the pent roof cylinder head is shown in figure 4. There are two independent variables which are the inclination angle α and the off-set fraction $\xi = c/a$. The cylinder head is divided into two parts; the major section, shown on the right, is inclined at an angle α to the normal to the cylinder axis. It is regarded as the inlet side of the head and it projects as the larger part of an ellipse with major and minor axes of $2a$ and $2b$ respectively.

The smaller section on the left is inclined at an angle ψ to the normal to the axis of the cylinder; this is regarded as the exhaust side of the head. The line where the two sections meet is off-set from the axis of the cylinder by an amount c measured along the major axis of inlet ellipse. The two inclination angles α and ψ and the off-set fraction $\xi = c/a$ are related by

$$\tan \psi = \left(\frac{1 + \xi}{1 - \xi} \right) \tan \alpha \tag{1}$$

It is interesting and important to note that the two off-set fractions c/a and c_l/a_l are numerically equal. Thus

$$\xi = c/a = c \cos \alpha / b = |c_L/a_L| = |c_L \cos \psi / b| \tag{2}$$

where c and c_L are measured along the inclined planes, as shown in figure 4. It is convenient to regard the off-set fraction as positive for the inlet, the right hand side, and negative for the exhaust, the left-hand side. The arrangements for the valves are shown in figures 5 and 6.

One inlet valve and one exhaust valve

When only one inlet valve is used with one exhaust valve, the generic area of the inlet valve is

$$\pi r^2 = \pi \left(\frac{a + c}{2} \right)^2$$

and the generic area ratio is

$$\Omega_i = \left(\frac{a + c}{R} \right)^2 \tag{3}$$

The generic area ratio of one exhaust valve is

$$\Omega_e = \left(\frac{a_L - c_L}{R} \right)^2 \tag{4}$$

Other equations are given in the Appendix and the calculated results shown in figure 7.

Two inlet valves and two exhaust valves

The arrangement for the two inlet valves is shown in figure 5. The generic circle for the valve is tangential to the ellipse at P, the abscissa at B and to the line of intersection of the two sections at A. The angle ϕ is the parameter associated with the auxialliary circle. The relevant equations for the four valve case are given in the Appendix. These equations have been solved numerically and the resutls for both inlet and exhaust valves are shown in figure 8.

Three inlet valves and three exhaust valves

The case of three inlet valves is shown in figure 6. The valve generic circles are tangential to one another in addition to the tangent points from the previous case and to the ellipse at C. Again, the relevant equations are solved numerically and given in the appendix. The results for the inlet and exhaust valves are shown in figure 9.

Other valve combinations

The above method of analysis enables the evaluation of other combinations of valves to be made. This is illustrated using the five valve case, namely three inlet valves and two exhaust valves. It is shown by points X on figure 9, Y on figure 8 and results in Table 1. This technique is possible because for a given inclination angle α and fractional offset for the larger section ξ the corresponding fractional offset for the smaller section is numerically the same but opposite in sign to that of the inlet section.

On the diagrams shown as figure 7, 8 and 9 the contours are shown for constant α over a range of off-set fractions. It should be appreciated that along a given contour, α has a fixed value but the inclination angle Ψ varies as shown by equation 1. Hence, a given point on the inlet side $i.e.$ a given α and fractional off-set on one of figure 7,8 or 9 may be matched with the corresponding α and the negative of the off-set fraction on any one of figures 7 8 and 9. This principle is used in the remainder of the paper.

4. Discussion

The key ideas of the present paper are the extensions of the previous treatment of flat head multi-valve schemes to include pent roof designs with various amounts of off-set projections. The new treamtment allows different diameters for inlet and exhaust valves. However, for a given multi-valve design, all the inlet valves have the same size and all the exhaust valves have a common size which in general is different from that of the inlet valves. An important aspect of the comparison is the use of the generic area ratios. A number of relatively small factors have been neglected. They are

a. the difference due to flow direction between exahust and inlet valves

b. the mechanical and aerodynamic interference between adjacent valves and between valves and cylinder walls

c. the implied complexity of the required valve gear and associated valve pipework.

The datum value of the generic valve is that of a circle with a diameter equal to the cylinder radius. This is the same value used previously in ref [6]. The modulus of the off-set ratio for a given set of exhaust valves is equal to the off-set ratio for a suitable set of inlet valves. The basic effect of increase in the inclination angle α, and the off-set ratio is for all three valve arrangements. Figure 7,8 and 9, to increase the inlet area. But the basic efects of increasing the inclination angle Ψ and reducing the offset ratio is to increase the exhaust valve area.

The contours for various inclinations diverge with increases in off-set ratio for the cases of one and three valves per cylinder projection. In the case of two valves per cylinder the corresponding contours converge with increasing off-set ratios.

The largest area of inlet valves occurs on the three valve case, with the single valve case ranked second. The contours for the two-valve case converge with increasing off-set fraction because there is a reducing increase in the diameter of the valve generic circle as the inclination angle α is increased.

As the off-set is increased for the inlet valves, the inlet area increases and as expected there is a corresponding reduction in exhaust area. This reduction is most pronounced for the single exhaust valve case.

In order to assess the merits of different schemes on the basis of inlet and exahust areas simultaneously, some results have been plotted in a novel way on figure 10. Here the optimum domain is the top right hand region of the diagram. Two points stand out. First, the six valve head, three inlet, three exhaust valves case. Engines with 5 valves per cylinder were mentioned in the discussion paper (ref [2]) by Amann, Pearson and Wallace and reference was made to five valve engines by Ferrari, Yamaha, Audi and Bugatti.

Other combinations of valve arrangements may be identified on Figure 10 using the same technique as that used above for the five valve case, in order to find cases of different contour ratio Ω_i/Ω_e which may be suggested from engine simulation surveys.

Comments brought out by Wallace [2] are that multi-valve schemes have smaller diameter valves with smaller reciprocating mass and operation at higher engine speeds. They are also able to operate at higher valve lift ratios, an effect which is not brought out in the diagrams shown in this paper.

Finally, when ranking the advantages discussed above, the interference between adjacent valves and valves and cylinder walls should be addressed in an actual design. The interference is both mechanical and aerodynamic in nature.

5. Conclusions

The method of comparing the flow area of poppet valve schemes developed previously on a comparative basis for schemes with equal size inlet and exhaust valves and a flat cylinder head has been extended. The extensions introduced here have made the assessment much more realistic because it now includes pent roof designs with off-set projections for the valve planes. Assessments have been made for inlet valve projections up to 30° and offset ratios up to 20%.

The method has shown a gain of 98% on the inlet area for the three inlet valve case with 20% off set and $\alpha = 30^o$ and a gain of 31.7% on the exhaust area for three exhaust valves. This only reduces to 30% for the case of two exhaust valves.

When the offset is reduced to 10% for the three inlet valves and three exhaust valves case, the gain on the inlet area is still 76.5% but the gain on the exhaust area is 42.8%.

This paper, and particularly figure 10 and the above examples, have presented a most useful method of ranking the flow capacity to assist the design of multi-valve schemes. Finally a series of overall engine computer simulation studies are needed over a range of inlet to exhaust valve area ratios to determine the optimum values. When this is established the methods described in this paper can be used to select the valve numbers, section inclination angle and the off-set fraction required.

Appendix

Equations for generic area of one, two and three valves per projection surface shown on figure 4, figure 5 and figure 6. For inlet c/a is positive, $b/a = \cos\alpha$. For exhaust c/a values are negative and b_L/a_L is used in place of b/a where $b_L/a_L = \cos\psi$.

A.1 One valve per projection

$$\frac{r}{b} = \frac{1 + \frac{c}{a}}{2\frac{b}{a}} \tag{A1}$$

$$\Omega = 4\frac{r^2}{b^2} \tag{A2}$$

A.2 Two valves per projection (figure 5)

$$\tan\phi = \frac{b}{a}\tan\beta \tag{A3}$$

Equation solved by numerical methods

$$\frac{\frac{b}{a}\sin\phi}{\cos\phi + \frac{c}{a}} = \frac{1 + \sin\beta}{1 + \cos\beta} \tag{A4}$$

Radius ratio equation

$$\frac{r}{b} = \frac{\frac{c}{a} + \cos\phi}{\frac{b}{a}(1 + \cos\beta)} \tag{A5}$$

$$\Omega = 8\frac{r^2}{b^2} \tag{A6}$$

A.3 Three valves per projection (figure 6)

$$\tan\phi = \frac{b}{a}\tan\beta \tag{A7}$$

Equation solved by numerical method

$$\frac{1}{4}\left[\frac{\frac{b}{a}\sin\phi(1 + \cos\beta)}{\cos\phi + \frac{c}{a}} - \sin\beta\right]^2 + \left[\frac{(1 + \frac{c}{a})(1 + \cos\beta)}{2(\cos\phi + \frac{c}{a})} - 1\right]^2 = 1 \tag{A8}$$

Radius ratio equation

$$\frac{r}{b} = \frac{\sin\phi}{2\sin\gamma + \sin\beta} \tag{A9}$$

$$\Omega = 12\frac{r^2}{b^2} \tag{A10}$$

References

1. Heywood, J.B., Internal combustion engine fundamentals, 1998, pp 1-930, McGraw-Hill, New York

2. Amann, C.A., pp 407-408; Pearson, R.J., pp. 408-410; Wallace, F.J., pp 410-411. Discussion paper on reference 6. Proc Instn Mech Eng, Vol 211, Part D, Journal of Automotive Engineering 1997

3. Hosaka, T and Hamazaki, M., Development of the Variable Valve Timing and Lift (VTEC) Engine for the Honda NSX, The Fifth Autotechnologies Conference and Exhibition, Monte Carlo, Monaco (SAE p238) 23-25 January, 1991, paper 910008.

4. Setright, L.J.K., Some Unusual Engines, Mechanical Engineering Publications Limited, 1975, p66.

5. Stone, R., Introduction to Internal Combustion Engines, MacMillan Education Ltd., Hampshire and London, 1985, p319.

6. Woods, W.A., and Brown, P.G., Flow area of multiple poppet valves, Proc Inst Mech Engrs., Vol 210, Part D, Journal of Automobile Engineering, 1996, pp 347-351.

7. Woods, W.A., and Khan, S.R., An experimental study of flow through poppet valves, Proc Inst Mech Engnrs., Vol 180, Part 3N, 1965, pp 32-41.

Table 1 : Illustration of five valve head

On figure 9, point X, three inlet valves

$\alpha = 30^o$

$\xi = 0.2$

$\Omega_i = 1.98$

On figure 8, point Y, two exhaust valves

$\psi = 40.89^o$

$\xi = -0.20$

$\Omega_e = 1.2999$

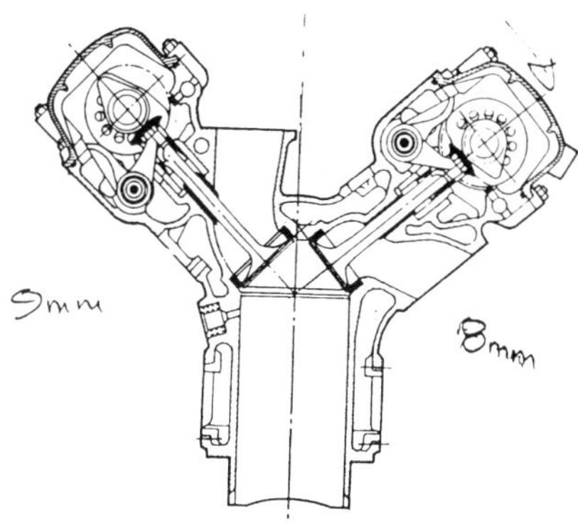

1955 MERCEDES-BENZ   M 196 / 110 ENGINE WITH PENT ROOF
COMBUSTION CHAMBER AND DESMODROMIC VALVE GEAR

FROM SETRIGHT REF 4   FIG 1

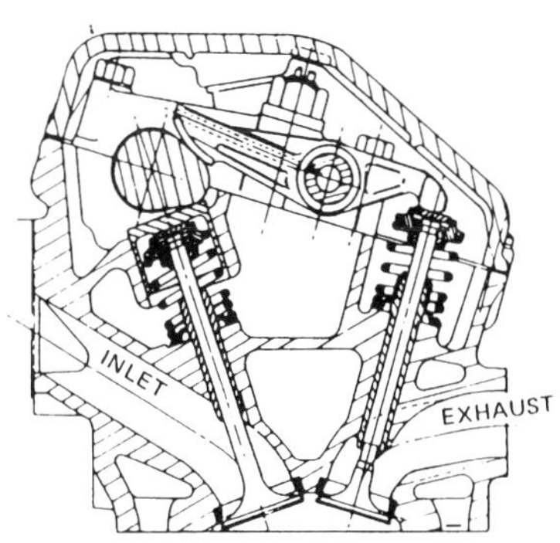

BRITISH LEYLAND 4 VALVE PENT ROOF HEAD DESIGN

FROM STONE (ref. 5)   FIG 2

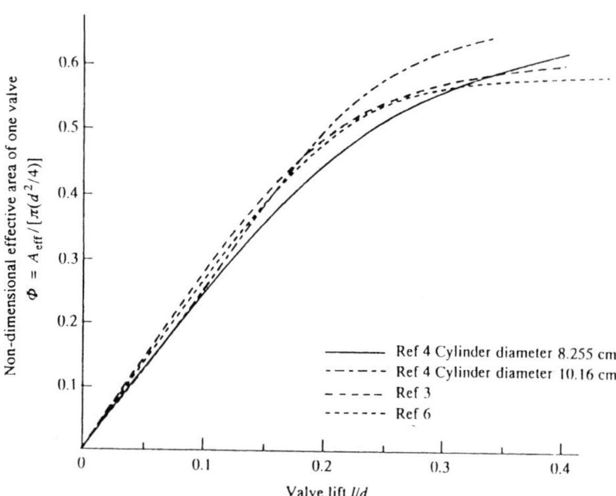

Fig. 1 Comparison of effective areas of poppet valves [adapted from reference (4)]

NON-DIMENSIONAL EFFECTIVE AREA OF POPPET VALVE AS A FUNCTION OF NON-DIMENSIONAL LIFT

adapted from Ref [7]   FIG 3

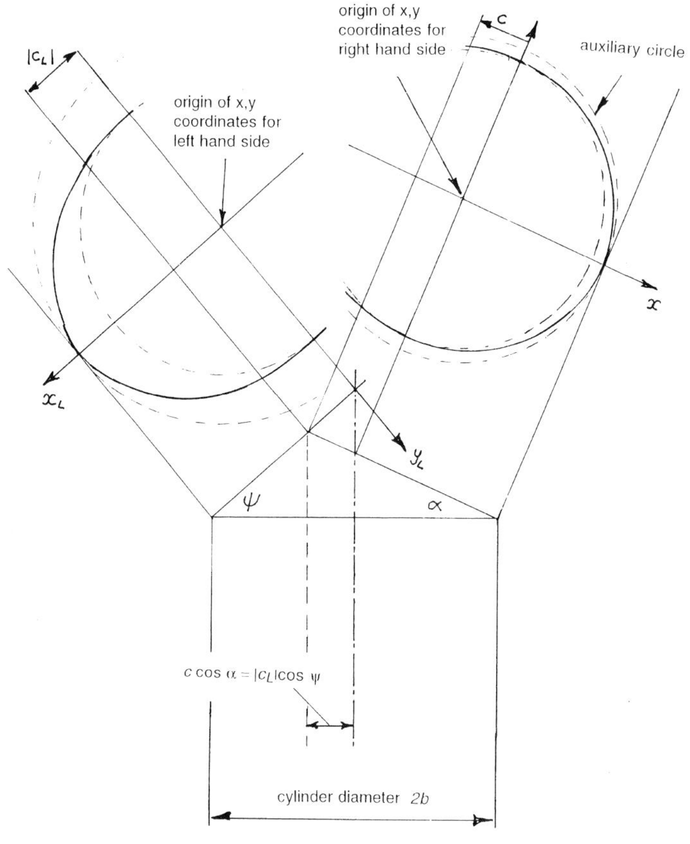

DIAGRAM SHOWING OFF-SET PENT ROOF WITH PROJECTIONS OF SURFACES

FIG 4

NOTES:

$(a+c) \sin \alpha = |a_L + c_L| \sin \psi$

$c \cos \alpha = |c_L| \cos \psi$

$b/a_L = \cos \psi$

$b/a = \cos \alpha$

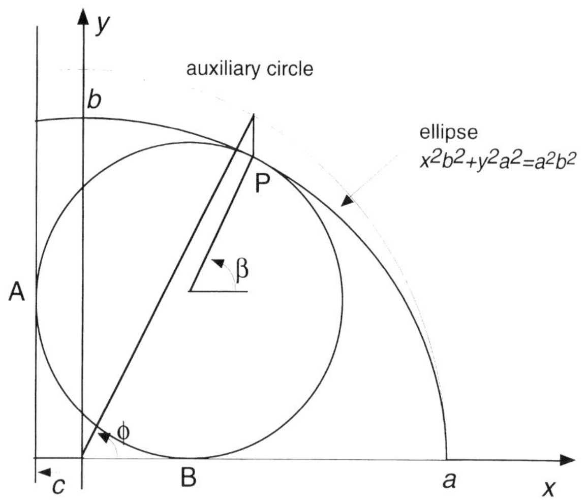

ARRANGEMENT FOR TWO INLET VALVES   FIG 5

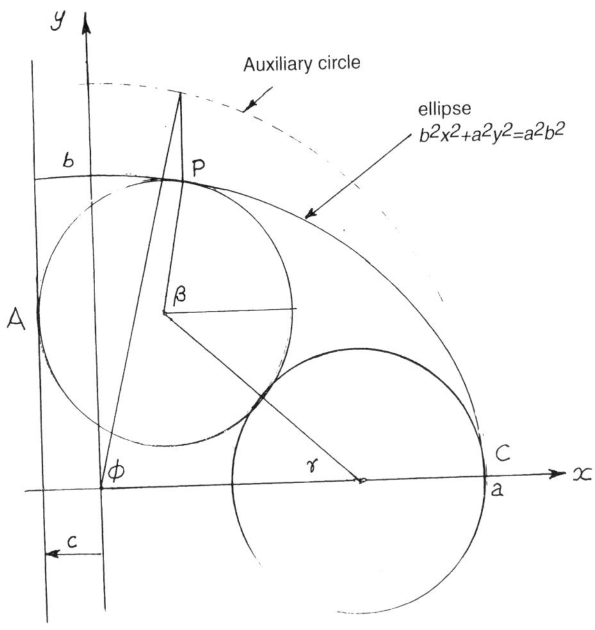

ARRANGEMENT FOR THREE INLET VALVES   FIG 6

FIG 7

Generic area ratios for two valve head

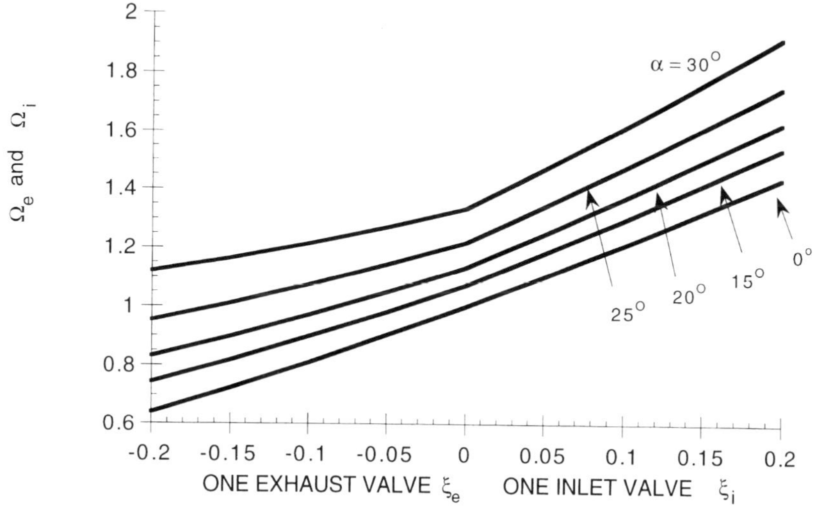

FIG 8

Generic area ratios for four valve head

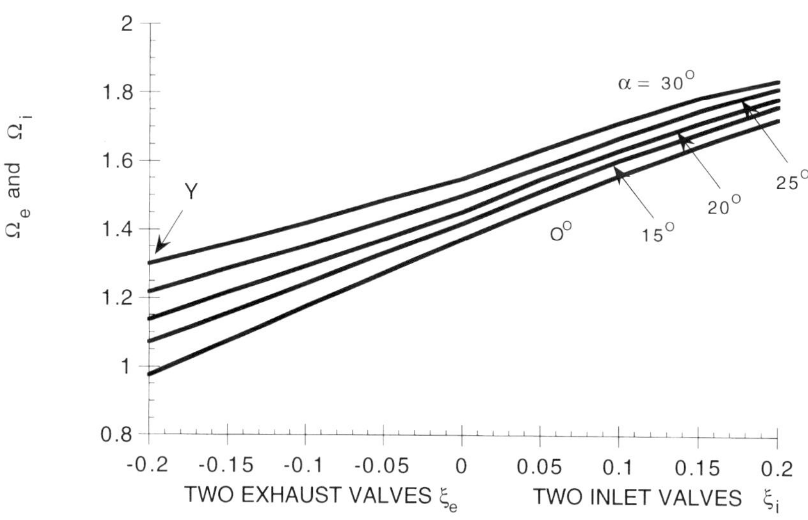

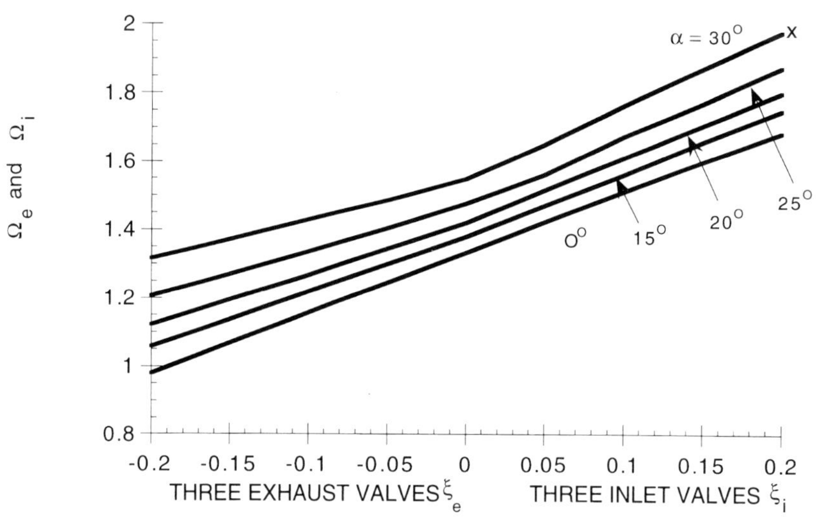

Generic area ratios for six valve head
FIG 9
Ω_e and Ω_i
2
1.8
1.6
1.4
1.2
1
0.8
$\alpha = 30^{O}$
x
O^{O}
15°
20°
25°
-0.2
-0.15
-0.1
-0.05
0
0.05
0.1
0.15
0.2
THREE EXHAUST VALVES ξ_e
THREE INLET VALVES ξ_i

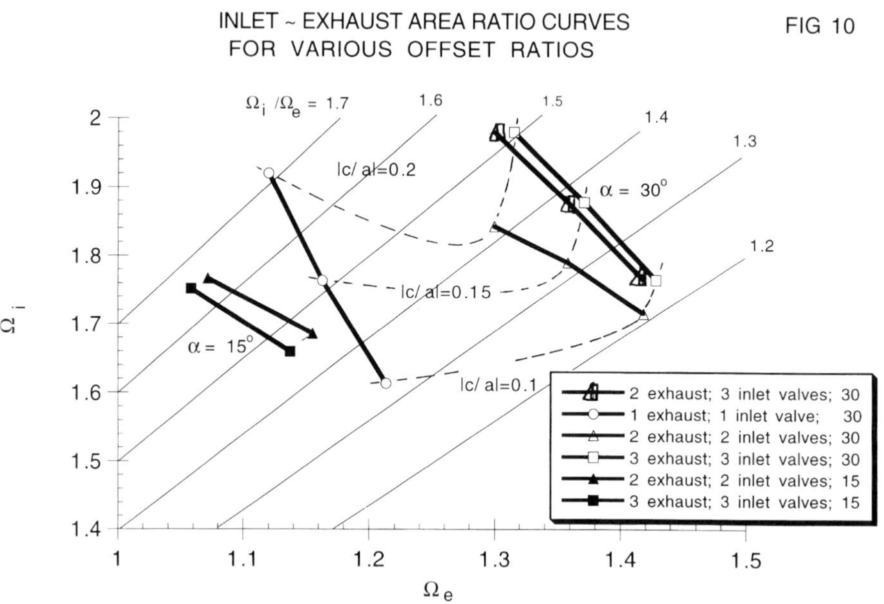

INLET ~ EXHAUST AREA RATIO CURVES
FOR VARIOUS OFFSET RATIOS
FIG 10
$\Omega_i / \Omega_e = 1.7$
1.6
1.5
1.4
1.3
1.2
2
1.9
1.8
1.7
1.6
1.5
1.4
Ω_i
lc/ al=0.2
lc/ al=0.15
lc/ al=0.1
$\alpha = 30^{O}$
$\alpha = 15^{O}$
2 exhaust; 3 inlet valves; 30
1 exhaust; 1 inlet valve;   30
2 exhaust; 2 inlet valves; 30
3 exhaust; 3 inlet valves; 30
2 exhaust; 2 inlet valves; 15
3 exhaust; 3 inlet valves; 15
1
1.1
1.2
1.3
1.4
1.5
Ω_e

S608/004/99

Modelling of transient flows in a four-valve pent-roof type engine with late intake valve closure

A Y BOKHARY
College of Engineering, King Abdul-Aziz University, Saudi Arabia
R J SAUNDERS
Department of Mechanical Engineering, University of Sheffield, UK

SYNOPSIS

Understanding of flow structures in the fuel-air mixture before ignition is needed to assess the combustion performance of spark ignition engines, particularly if the combination of swirl and tumble is likely to be modified by Late Intake Valve Closure (LIVC). This study has used the STAR-CD code with the k-ε turbulence model to simulate the relevant transient flows, after using some steady flow hot wire anemometer measurements to provide a validation of the code. A normal valve timing is compared with that for an LIVC mechanism in which one of the two inlet valve events is retarded.

At the end of induction both mean velocities and turbulence kinetic energies are greater for LIVC than for the normal timing. They also show a small net swirling motion in addition to the tumbling motion and the flow structure appears more complex. These effects account in part for the good combustion characteristics observed in an Otto-Atkinson cycle engine.

1. INTRODUCTION

This study arose from earlier experimental work on an engine with Late Inlet Valve Closure (LIVC) [1]. In that work engine load was reduced by progressive retardation of IVC into the compression stroke in such a way that the effective compression ratio was significantly reduced. It was anticipated that combustion quality might well deteriorate under these conditions as a result of both lower compression temperature and modified air motion, but no such deterioration was observed. Although this encouraging result may be attributable to improved mixture formation, the role of possible changes in air motion and turbulence was unknown and hard to predict. In particular the effect of a delayed opening of one of the two inlet valves was expected to affect the balance between tumble and swirl, and the resulting structure of turbulence and swirl at the end of compression.

Air motion inside the combustion chamber is the soul of engine performance. It is quite clear that the structure of the air motion and the level of turbulence kinetic energy have important effects on flame propagation through the air-fuel mixture and hence on the combustion of the entire mixture. Complete combustion of the air-fuel mixture inside the engine is an essential objective in order to meet continuing demands for reductions in fuel consumption and exhaust emissions. A number of both experimental and computational studies have investigated the various factors that might affect the structure of the air motion and enhance turbulence in an engine cylinder.

Intake system configuration is of primary importance, and Arcoumanis [2] showed experimentally the effect of different inlet port orientations on in-cylinder air motion, turbulence kinetic energy and the burning rate of the fuel. Although pure tumble generated more turbulence at ignition than combined tumble and swirl, it was nevertheless suggested that the existence of swirl motion may itself enhance the burning rate. The measurements of Jaffri [3] showed that asymmetry of the inlet port runner may cause different tumble ratios in different measurement planes, with almost zero swirl throughout. Similarly, flow visualisation [4] of in-cylinder flows and combustion had shown a breakdown of tumble motion well before TDC as an enhancement of the tumble intensity, which can reduce the initial burn period without significant overall increase in burn rate.

Computationally, Delhaye and Cousyn [5] simulated the tumbling structure expected in a four-valve pent-roof type combustion chamber design, and showed the effect of engine speed in preserving tumble motion during the compression stroke. Similarly, Jones and Junday [6] completed a full cycle simulation and showed the effect of inlet port design in intensifying the tumble motion.

1.1 Variable Valve Timing

Among the different factors that might affect in-cylinder air motion is the inlet valve timing, and inlet valve events may be expected to play a significant role in this structure. Variable valve timing (VVT) is a particularly topical concept, which is most commonly applied to the inlet valves, particularly to improve performance of the engine but also to reduce emissions and fuel consumption. Early intake valve opening (EIVO) has been used to improve volumetric efficiency, with inlet valve openings as early as 55° BTDC [7]. Early intake valve closure (EIVC) can be used for load control by restricting the end of induction, and at low speed it may also be used to improve the torque curve [8].

Similarly, LIVC may be used to enhance the torque curve at high engine speed, whereas at lower speeds it will cause some of the charge to flow back into the inlet and hence act as a method of load control. Improvements in part load efficiency with LIVC have been demonstrated over a range of LIVC up to 120°ABDC [1,9,10,11]. As engine speed is increased at given LIVC, back flow into the inlet manifold is progressively reduced until the pressure in the cylinder at IVC becomes equal to that in the inlet manifold. Beyond this speed, LIVC can give higher volumetric efficiency and greater torque. These advantages compensate for an effective reduction in the compression ratio. At very high speeds, LIVC may show lower brake torque than at mid engine speeds because of increases in fluid friction along the walls and valve surfaces, and hence increased pumping loss [7,8].

The objective of this paper is to use computational fluid dynamics (CFD) to study the effect of LIVC on the in-cylinder flow structure and turbulence kinetic energy generation in a four-valve pent roof type engine. The LIVC system used here follows that implemented in [1], in which the two inlet valves retain the same lift event design but one of them is substantially retarded from the standard timing of the other one. The CFD code used is STAR-CD, both for the generation of a suitably complex mesh including two independently moving valves and for the transient numerical solution of the air flow through the moving mesh so created.

2.   CODE VALIDATION IN STEADY FLOW

Firstly a comparison was made between measurements from a simple steady flow experiment and a corresponding steady flow simulation using STAR-CD with the k-ε turbulence model, in order to provide some form of validation for typical flow conditions in the engine. A computational mesh was set up for an open cylinder with flow through two valves open at fixed positions, corresponding with the geometry of an actual four-valve cylinder head on a flow bench and bolted to an open perspex cylinder. The exhaust valves were kept closed, and air flowed out of the end of the cylinder to give a flow distribution somewhat similar to that during an induction stroke. Velocities were measured in turbulent steady flow by hot wire anemometer (HWA) over a range of accessible locations in the cylinder and combustion chamber. Advantage was taken of the design symmetry of the engine cylinder in question so that both measurements and simulation could be confined to one half of the flow system, bounded by a plane of symmetry. This also however required that the two inlet valves should be open at the same lift. The experiment was performed for two valve lifts, half and fully open.

2.1   Hot Wire Anemometer Measurements
The HWA is a thermal transducer of very low inertia, and was selected as a simple method of obtaining both mean and turbulent velocity components. When an electric current passes through the fine wire sensor, the power dissipated measures the heat transfer from the wire and hence the fluid velocity at the wire. When operated at constant temperature, the anemometer voltage can be directly correlated with the fluid velocity normal to the wire [12]. In a random flow field however, a strategy is required either to decompose the signals from multiple sensor orientations into velocity components in the three principal directions, or to obtain information about the angle of flow before a measurement is recorded. The former strategy is effective when turbulence levels are not too high relative to the mean velocity, but in our case there are regions of high turbulence with low mean velocity, for which the second method was considered more practicable.

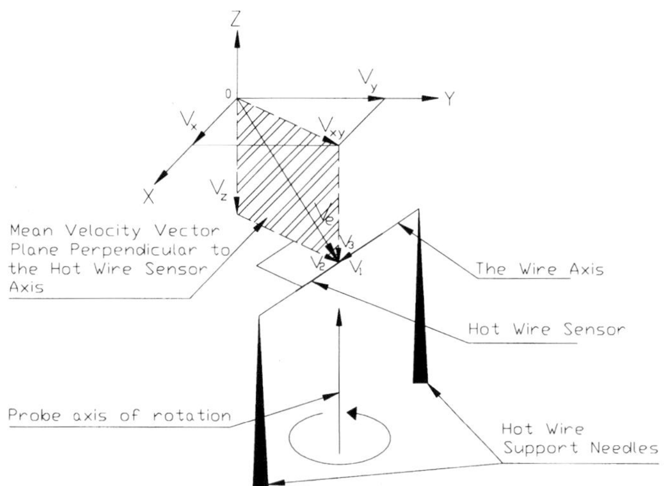

Figure 1 Sketch of the HWA sensor when exposed to a flow [13].

A single straight hot wire probe is used for the internal flow measurements by rotating the probe about the axis of the probe holder until a maximum signal appears on a digital voltmeter. This implies that the mean velocity vector is perpendicular to the hot wire sensor. Figure 1 shows how a single hot wire probe can detect the mean velocity magnitude when the sensor is set perpendicular to the plane of the velocity vector, but it will not give the precise direction of the velocity vector. The following relations show how the mean velocity magnitude detected by the hot wire sensor is related to velocity components in the fixed x,y,z directions and those relative to the wire axis [13].

$$V_e = \sqrt{V_x^2 + V_y^2 + V_z^2} \qquad \text{and also} \qquad V_e = \sqrt{V_1^2 + V_2^2 + V_3^2}$$

but since $V_1 = 0$, $\quad V_2 = V_{xy} = \sqrt{V_x^2 + V_y^2}$ and $V_3 = V_z$, then

$$V_e = \sqrt{V_{xy}^2 + V_z^2}$$

Turbulence was assumed to be isotropic, so that the fluctuating velocity component was taken to represent the turbulent kinetic energy k in only two of the three principal directions, and a correction factor of 1.5 was therefore applied to k.

Figure 2 shows a schematic of the flow bench and measurement system. Air from the compressor flows through the orifice plate into the cylinder through the inlet port and valves, which are set at the desired lift by micrometers. The hot wire probe is accurately positioned at a desired point and rotated about its vertical axis until a maximum mean voltage is observed. The dynamic HWA signal is then sampled at a suitable high frequency by a PC with computer-scope card and processed to yield the mean velocity and turbulent intensity magnitudes for that location. By this means data can be assembled covering a whole range of locations through the combustion chamber and cylinder.

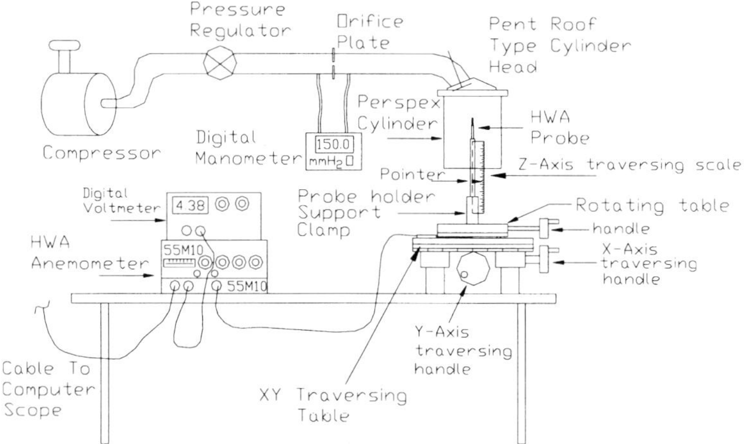

Figure 2   **Sketch of the experimental apparatus [13].**

2.2   Steady Flow Simulation

Two similar mesh structures of about 32,000 cells each were created for the steady flow simulation (fig 3) to represent two settings of inlet valve lift: 10 mm (fully open) and 5 mm (half open). This mesh was designed to incorporate geometrical structures which might particularly be required in a moving mesh in order to replicate any mesh dependent effects. An inlet boundary condition was set with a constant velocity of 24 m/s at the inlet port, corresponding to the measured flow. The bottom of the open cylinder was set as an outlet boundary condition. A symmetry boundary condition was set for the diametral plane between the two inlet valves. The wall default boundary condition was set for other unspecified surfaces. A number of vertices (monitoring nodes) were placed inside the mesh at locations corresponding to those used for the hot wire measurements, to allow precise comparisons to be made.

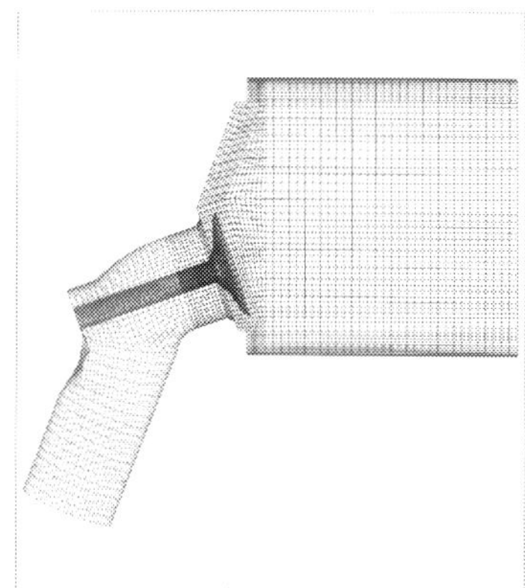

Figure 3   **Mesh for steady flow (10 mm valve lift)**

Examples of such comparisons between the measured and simulated results are shown in fig 4, covering both mean velocity and turbulence magnitudes for various lifts and locations. Agreement was generally good for mean velocities and moderate for turbulence. There were certainly some significant discrepancies, some of which reflect no more than slight displacement of a vortex structure while others indicate important overall differences in the flow structure. The measurements often suggest steeper local gradients of mean velocity than the simulation, and this may be attributed partly to limitations of the hot wire technique as well as to those of the k-ε turbulence model. Nevertheless the scale of agreement was regarded as giving sufficient encouragement to proceed with the full transient model.

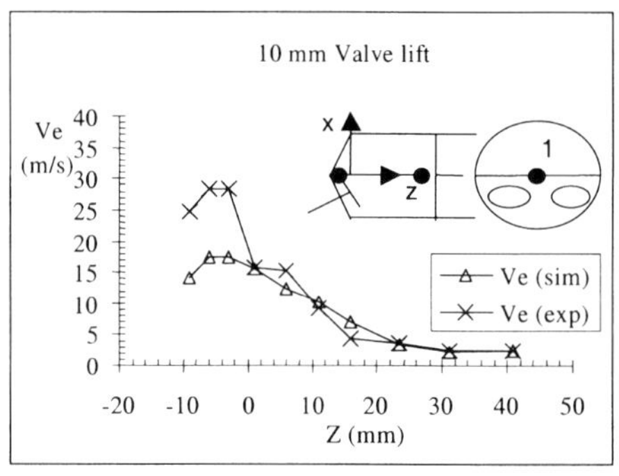

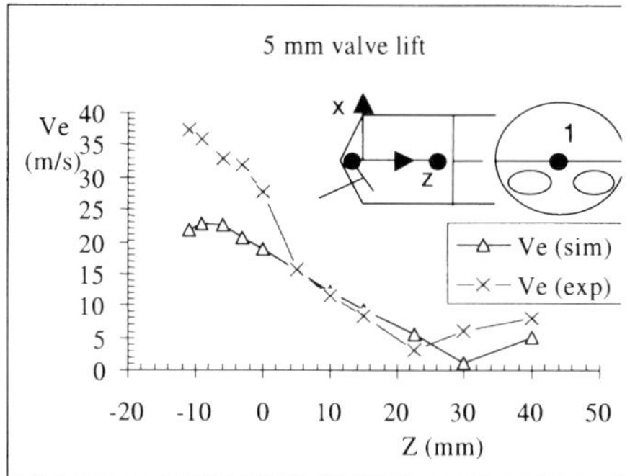

Figure 4(a)   Mean Velocities along the cylinder axis

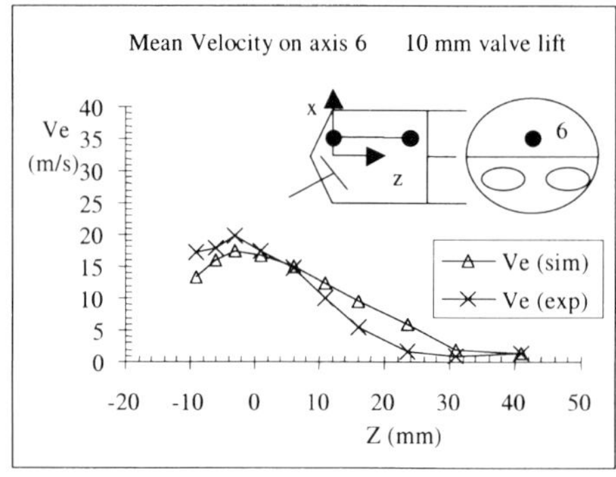

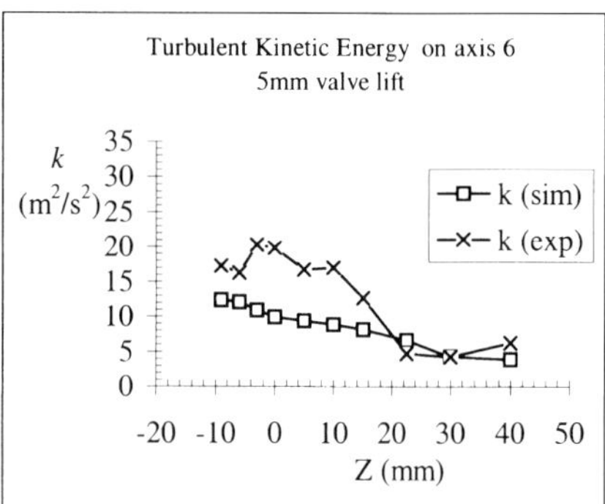

Figure 4(b)   Axis on exhaust side

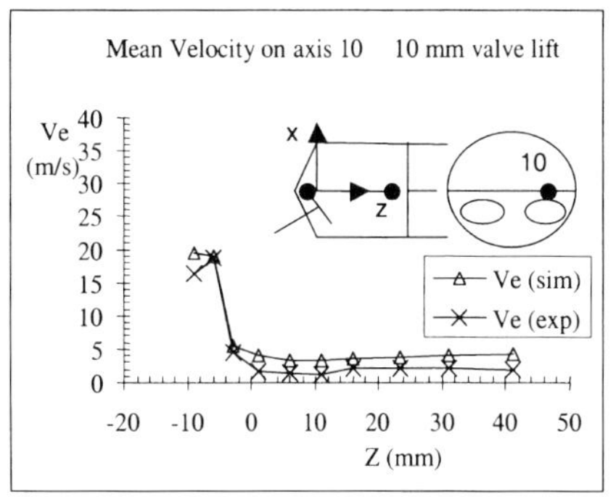

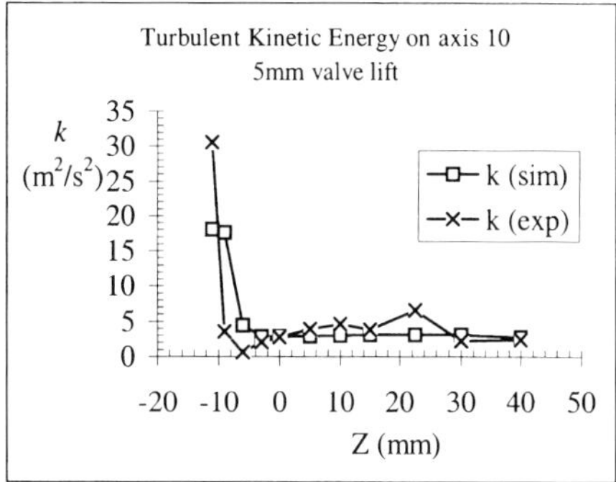

Figure 4(c)   Axis through jet from inlet valve

3.   SIMULATED FLOWS IN AN OPERATING ENGINE

Although the first mesh had included structures thought to be suitable for mesh movement, considerable mesh development was found to be necessary. Initially a new symmetrical (half cylinder) mesh was created with about 76,000 cells and a moving mesh design, to simulate the operation of the conventional engine with a moving piston and the two inlet valves moving in tandem. A second mesh of about 153,000 cells, which incorporated the first mesh with its own mirror image, was subsequently used with independent movement of the two valves to simulate LIVC operation (fig 5). As previously, a symmetry plane boundary condition was used in the first case, but otherwise the boundary conditions were the same for both cases. A constant pressure boundary condition was applied to the inlet port entrance, and some special 'attachment' boundaries were placed around the valve curtains to ensure flow continuity through the valves during valve motion. In order to avoid both exhaust valve overlap effects and a meshing problem at minimum cylinder volume, a steady flow solution was generated to correspond with the geometry and piston velocity a few degrees after TDC, and this file was then applied as the initial conditions in the cylinder.

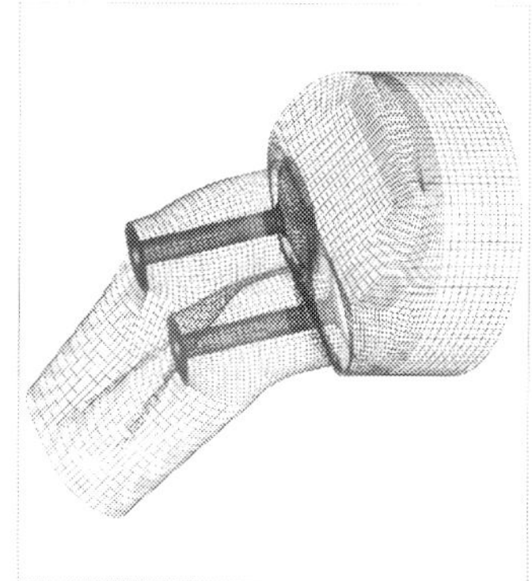

Figure 5   Full mesh for LIVC at 66°ATDC

Engine type	4 cylinder with 4 valves per cylinder pent roof type cylinder head
Engine Speed	1000 RPM
Bore and Stroke	86 mm
Inlet valve head dia.	34 mm
First valve: IVO (normal) IVC	2° BTDC 61° ABDC
Second valve: IVO IVC	28° ATDC 91° ABDC

Table 1 Engine specifications and inlet valve events.

Engine data and valve timings for the results to be shown are given in table 1. The conventional engine uses the timing of the first inlet valve only. Simulation results are presented for one case only of LIVC, such that the second valve event is uniformly retarded by 30° crank angle after the first. This represents a condition where LIVC has been shown to provide significant efficiency improvement at part load [1,11]. Larger timing differences may be expected to have more dramatic effects, but in all cases the most significant effect on air motion is likely to be that due to the late opening of the second valve. This results in higher air velocities through the valves early in the induction stroke and also different flows through each valve.

3.1   Flow structure comparisons

A typical flow structure during induction, in the plane of symmetry between the inlet valves, is shown in figure 6, indicating a low tumble system very similar to other published data (e.g. [14]). Figure 7 gives corresponding contours of turbulence kinetic energy. The direct influences of the valve flows are evident.

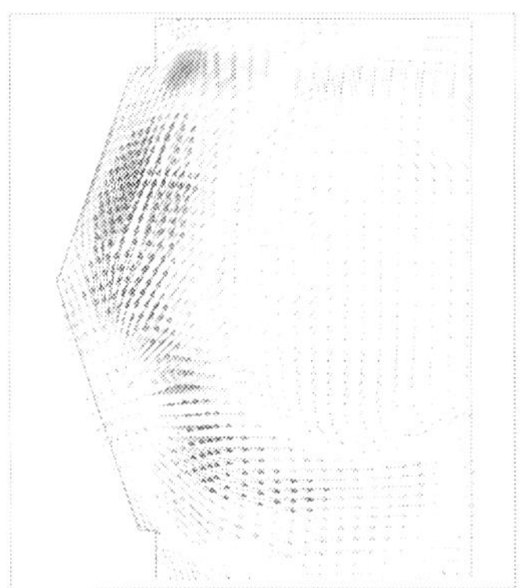

Figure 6   Mean velocity vectors (max 22 m/s)

Symmetry plane between valves, standard engine, mid induction stroke

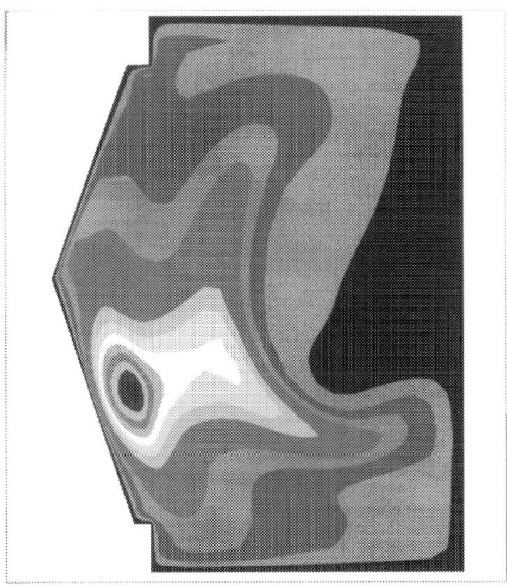

Figure 7   Turbulence energy (max 59 m^2 s^{-2})

Symmetry plane between valves, standard engine, mid induction stroke

Figures 8 and 9 compare the standard and LIVC cases on the basis of mean velocity structures in the plane of one of the valves.   LIVC gives a more extended jet from the valve at mid induction (fig 8), and a considerably more complex vortex structure at BDC (fig 9).

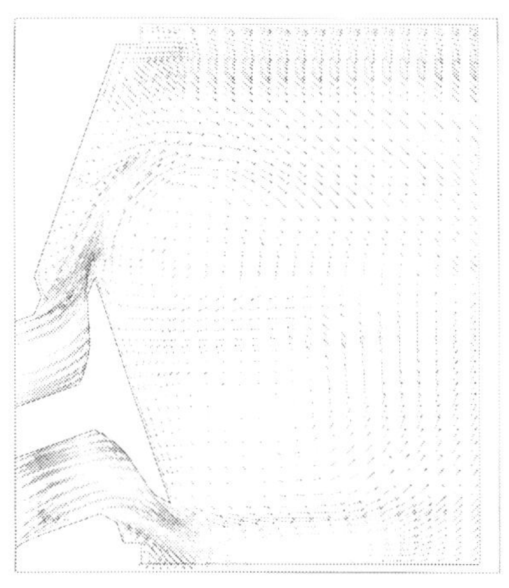

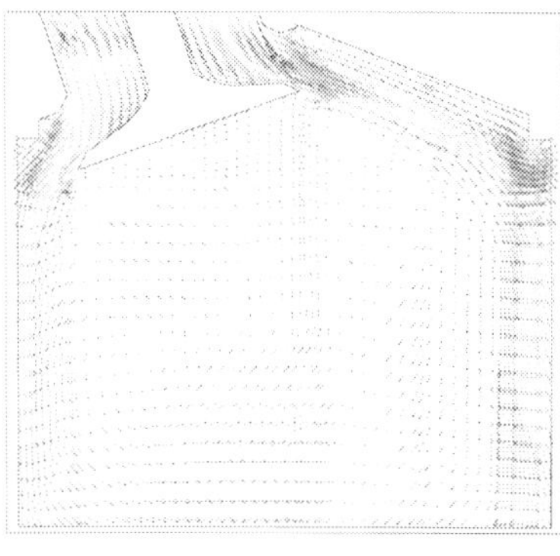

8(a) Standard engine (max 30 m/s)

8(b) LIVC (plane through first inlet valve, max 36 m/s)

Figure 8   Mean velocity vectors at mid induction

9(a) Standard engine	9(b) LIVC (plane through second inlet valve)

Figure 9   Mean velocity vectors at BDC induction (max 8 m/s)

Figure 10 compares turbulence energies for the same cases as fig 9, showing corresponding effects.   In an orthogonal plane including both inlet valves, figure 11 shows the flow asymmetry effects near BDC caused by LIVC in terms of both parameters.

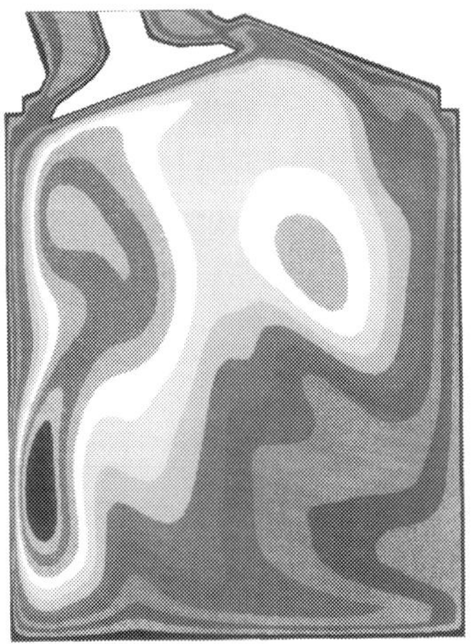

10(a) Standard engine (max 6 m^2s^{-2})	10(b) LIVC (plane through second inlet valve max 8 m^2/s^{-2})

Figure 10   Turbulence energy contours at BDC induction

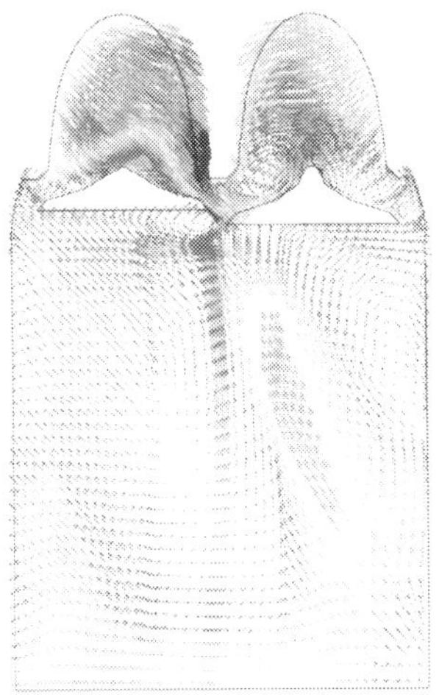 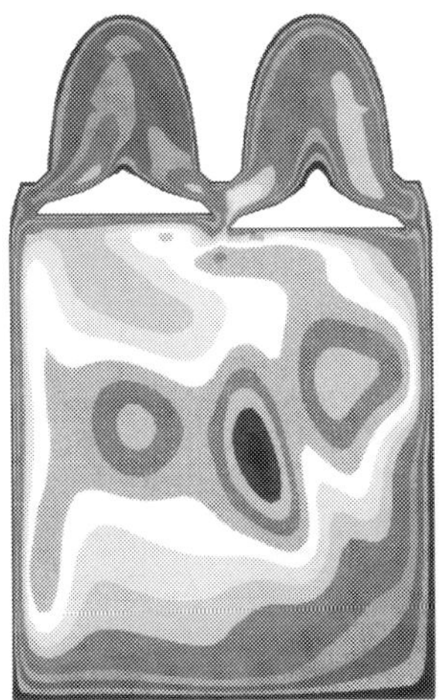

11(a)   Mean Velocity vectors (max 10 m/s)   11(b)   Turbulence energy (max 7 m^2/s^{-2})

Figure 11   Flow structures for engine with LIVC, end of induction stroke

Plots of these parameters along an axis allow a more compact form of comparison (figs 12,13), but also need to be viewed in conjunction with the wider picture. Figure 12 compares the two cases along the cylinder axis for a range of engine positions, whereas figure 13 shows the effects of asymmetry at the cylinder edge caused by LIVC compared with the standard case. Both mean velocity and turbulence appear to have decayed substantially by BDC on these axes, in spite of the substantial differences previously noted, but this is consistent with the low tumble design.

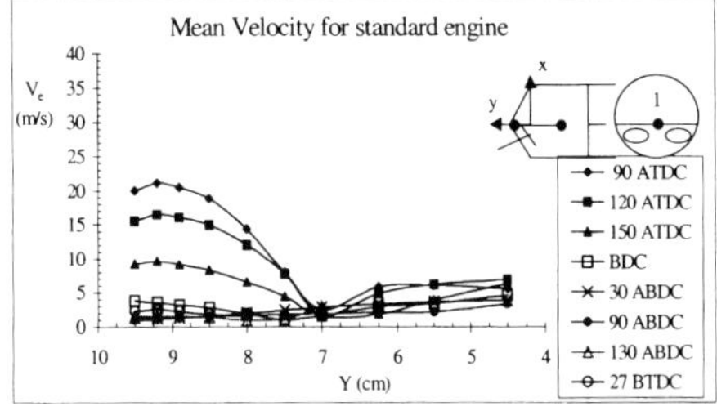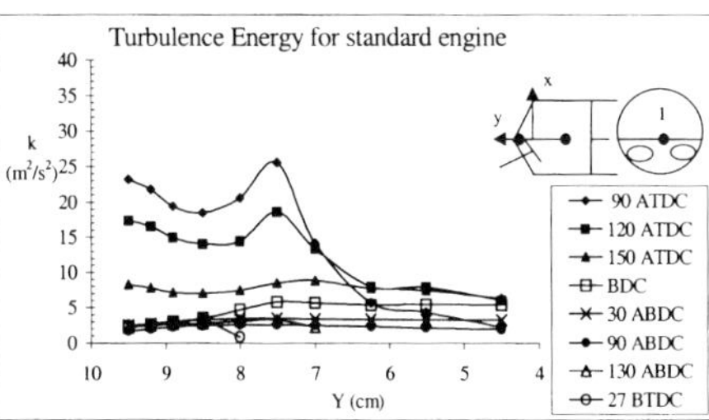

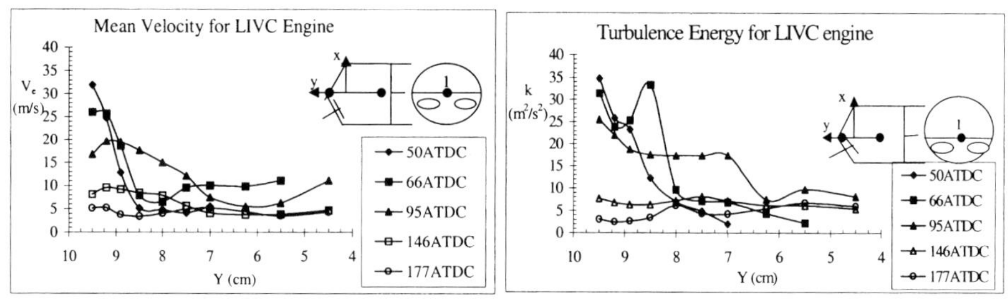

Figure 12   **Flow data on the cylinder axis**

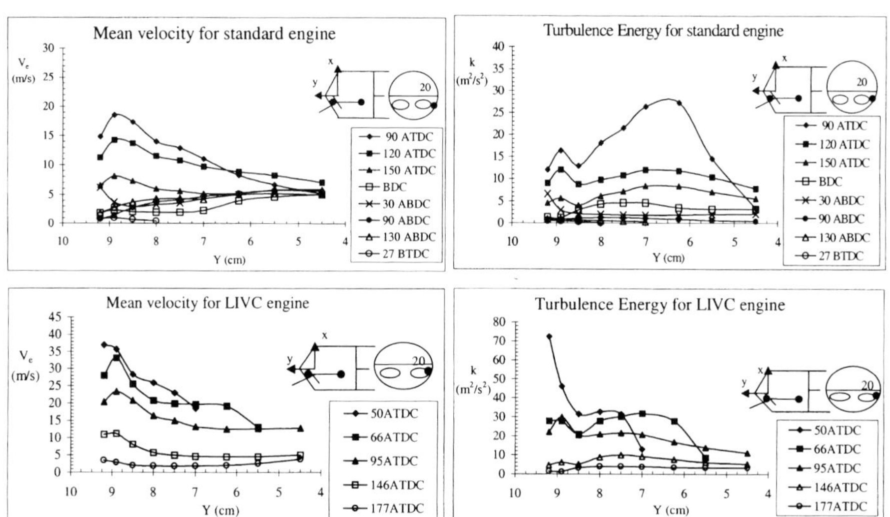

Figure 13(a)   **Flow data at cylinder edge (first inlet valve side)**

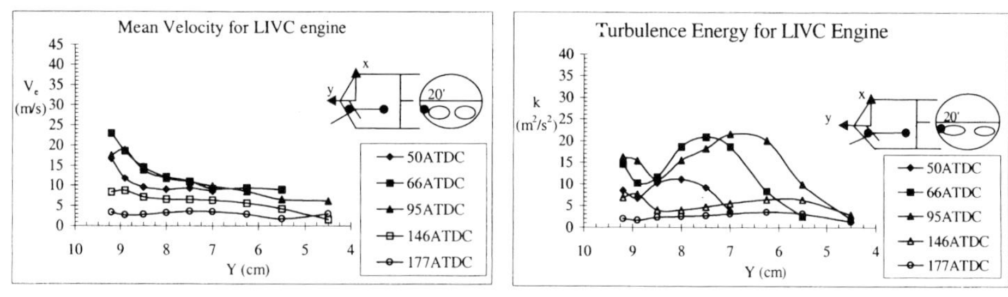

Figure 13(b)   **Flow data at cylinder edge (second inlet valve side)**

3.2   Discussion

The object of the study was to gain some insight about the effect of LIVC on in-cylinder air motion, with particular reference to conditions which would benefit combustion.   There is clear evidence that a LIVC system which uses a phase difference between the two inlet valves will generate both higher air velocities at the valves and a more complex vortex structure in the cylinder due to asymmetrical flow effects. As a result there is increased air motion and higher turbulence at the end of induction. The structure is most evident as separate tumbling vortices, with a small net increase in swirl. Because the intensity of tumble is not large for the standard engine, these effects may make a significant difference. A corresponding increase in turbulence intensity at the end of the compression stroke can be confidently predicted and hence an enhancement of the combustion process. The effect of LIVC on a high tumble design cannot be predicted from this study, but is expected to be relatively less favourable.

It is rather more difficult to quantify these effects for a number of reasons. Firstly, there are inevitably uncertainties attached to the use of the CFD code. The precision which can be obtained with the k-ε turbulence model for this type of flow is well known to be severely limited and this has been further demonstrated by the attempt at steady flow validation reported above. No significant alternative was readily available at the inception of this study, in view of the magnitude of the problem and the resources available. Also, it has not been possible to test the mesh or time step resolutions used for the same reason. Nevertheless a useful qualitative output has been obtained. Secondly, it has only been possible to cover a limited range of operating conditions, notably only one engine speed and one LIVC timing. Although this gives a good indication for real part load operating conditions, results at higher engine speeds are required before more general conclusions can be drawn.

4.   CONCLUSION

Use of the STAR-CD constant volume type CFD code with the k-ε turbulence model has yielded significant insight into in-cylinder air motion in a four-valve pent roof spark ignition engine, in which a realistic design of LIVC operation is implemented. The good combustion performance previously observed experimentally in such an engine can be partly attributed to the increases in air velocities and turbulence levels indicated by this study.

REFERENCES

1. Blakey, S.C., Saunders, R.J., Ma, T.H. and Chopra, A. "A design and experimental study of an Otto Atkinson cycle engine using late inlet valve closing" SAE 910451

2. Arcoumanis, C. et al, "Tumbling Motion: A mechanism for turbulence enhancement in spark-ignition engines," SAE 900060.

3. Jaffri, K. et al, "Tumble and swirl quantification within a motored four-valve SI engine cylinder based on 3-D LDV measurements," SAE 970792.

4. Kudou, H. et al, "A study about in-cylinder flow and combustion in a 4-valve S. I. Engine," SAE 920574.

5. Delhaye, B. and Cousyn, B., " Computation of Flow and Combustion in Spark Ignition Engine and Comparison with Experiment," SAE 961960.

6. Jones, P. and Junday, J., "Full Cycle Computational Fluid Dynamics Calculations in a Motored Four Valve Pent Roof Combustion Chamber and Comparison with Experiment," SAE 950286.

7. Assanis, D. N. and Polishak, M., "Valve Event Optimisation in a Spark Ignition Engine," J. Eng. Gas Turbines and Power, Vol. 113, pp 341-347, Jul. 1990.

8. Demmelbauer-Ebner, W., Dachs, A. and Lenz, H. P., "Variable Valve Actuation Systems for the Optimisation of Engine Torque," SAE 910447.

9. Saunders, R. J. and Rabia, S. M., "Part Load Efficiency in Gasoline engines," IMechE Seminar 'Practical Limits of Efficiency of Engines', pp. 55- 62, 12 Nov. 1986.

10. Saunders, R.J. and Abdul-Wahab, E.A., "Variable valve closure timing for load control and Otto Atkinson cycle engine" SAE 890677

11. O'Flynn, G.T., Saunders, R.J., and Ma, T.H. "Combustion Characteristics of an Otto-Atkinson Engine using Late Inlet Valve Closing and Multi-Point Electronic Fuel Injection" XXIV FISITA Congress, London 1992 paper 925107. I.Mech.E. C389/041 pp 329-338

12. Perry, A. E., "Hot Wire Anemometry," *Clarendon Press*, Oxford, 1982.

13. Bokhary, A. Y., "CFD Modelling of the Flow through a 4 Valve I. C. Engine with Late Intake Valve Closure," PhD Thesis, University of Sheffield, October 1998.

14. Khalighi, B., et al, "Computation and measurement of flow and combustion in a four-valve engine with intake variations" SAE 950287

Investigation of inlet port design on engine swirl using orthogonal array experimentation and computational fluid dynamics

A BRIGNALL
Ford Motor Company, Basildon, UK
Z M JIN
Department of Mechanical and Medical Engineering, University of Bradford, UK

This study represents a modern approach to quality of design for inlet ports in order to meet the increasingly stringent emissions legislation. Quality techniques such as a function tree and design of experiment (DOE) with orthogonal array testing are employed. Both experimental flow rig measurements and computational fluid dynamics (CFD) flow prediction are used to study the port performance in terms of swirl and flow coefficient. It has been found that the most important parameters are radius and wall position of the lip feature. Suggestions for minimising variation of port performance have been discussed.

1. INTRODUCTION

Implementation of stricter conformity of vehicle emissions requires a more focused approach to quality of design. One feature of the performance and emission system of direct injection (DI) diesel engine is the swirling motion of the inlet air charge around the cylinder axis. This swirl motion interacts with the fuel spray injected and dictates the nature of fuel combustion in the engine. Most features of this system such as fuel injector, valve seats, injector position and combustion bowl in the piston are machined to tight tolerances to ensure efficient combustion. The inlet port, which determines swirl, is cast into the cylinder head on mass produced engines. This research is based on the modern approach to quality engineering, which is meeting the customers needs.

For an inlet port the customers include the performance system engineers and manufacturers of the cylinder head, which includes the inlet port form. Many studies reported in the literature have revealed the port characteristics that give a good performance but few have focused on a port that can be made consistently for hundreds of thousands of engines per year.

Modern diesel engines adopt multi-valve approach as a means of achieving the increasingly low emission standards (1,2). This aspect of multi-valve engine layout is also well documented though little is published into the effect of interactions between port flows in order to determine the overall swirling motion. The combustion process is a customer of the swirl generating ability of the inlet ports, the swirl quality it receives is considered in this paper. Specific objectives of the present study are;

- The interaction between the ports.
- The design features of the ports.

2. METHODOLOGY

The role of an inlet port was analysed by the function tree technique to define the parameters which contribute to the port performance, and the relevant section is shown in Figure 1.

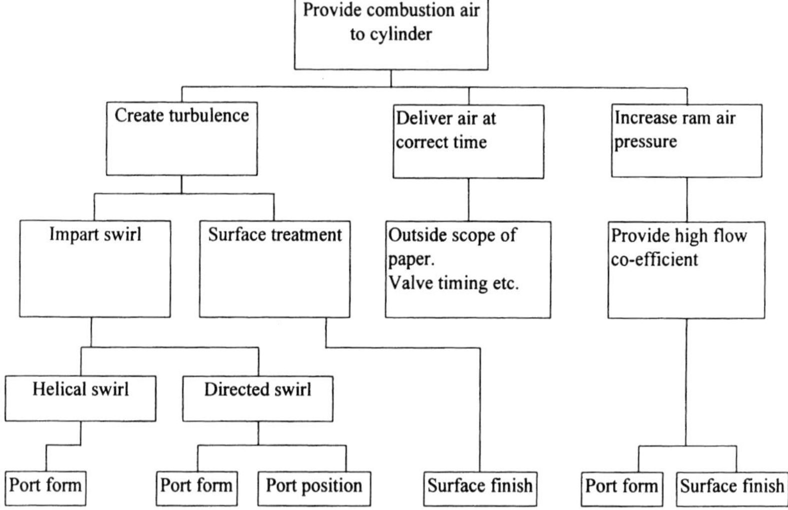

Fig.1 Diesel engine inlet port function tree

In Figure 1, the function of the port is described at the top level of the diagram and the steps how the function is achieved cascades down. A check on the rationale is tested by questioning why the action at one level should support the next level up.

2.1 Port interaction

The interaction between two inlet ports in providing air to the same cylinder was investigated on a sample of prototype multi-valve cylinder heads for direct injection diesel engines. The commonly used parameters for port performance are, primarily, swirl and, secondary, flow efficiency. These are usually determined at steady state flow conditions by a vane anemometer rig or an impulse reaction rig (3), where air is drawn through the inlet port and cylinder assembly and the speed of rotation of the air in the cylinder is measured.

Swirl performance of prototype multi-valve cylinder heads was investigated using a vane anemometer rig. The air swirl measured was then converted into a dimensionless ratio, defined as the swirl ratio, for convenience (Appendix 1). Port arrangement is shown in Figure 2 and the interaction of swirl performance is shown in Table 1.

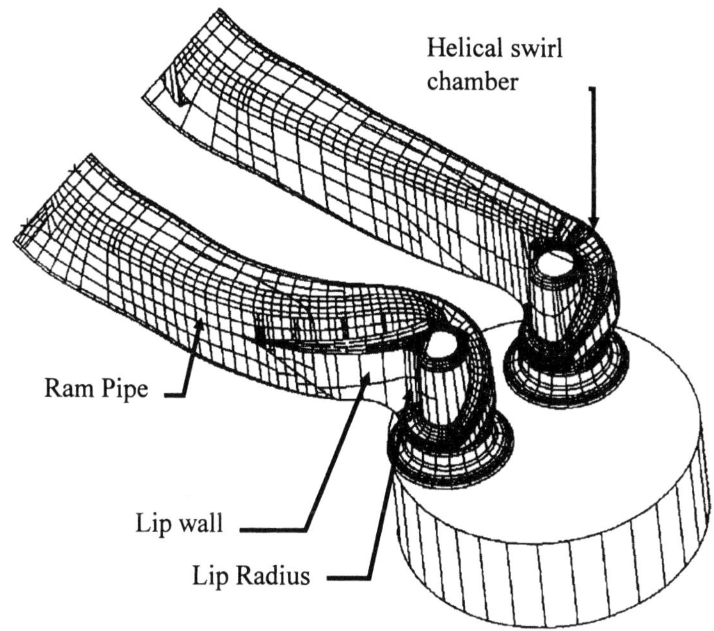

Fig. 2 Twin inlet port layout

Table 1 Variation in swirl ratio of a sample of prototype inlet ports

Feature	Primary port	Secondary port	Combined ports
Mean swirl ratio	3.52	1.58	1.85
Standard Deviation, s	.129	.089	.050
3s limit as percentage of mean swirl	11.0	16.9	8.2
Estimated % variation both ports open*			15.7

Sample size; 64 cylinder readings

* Using standard formula for combining standard deviations; $\sigma_b{}^2 = \sqrt{\sigma_p{}^2 + \sigma_s{}^2}$

These results presented in Table 1 show the swirl ratio for the combined ports to be less sensitive to variation than the individual ports and much lower than the estimated value from combining the individual port variation. It is generally recognised that the performance of a primary port is of most importance as the secondary port is deactivated at some operating conditions. The primary port is usually a helical swirl port in which the swirl is generated by the port roof having a helical form around the valve stem.

2.2 Helical swirl port

A list of features that affect port performance was brainstormed and a large number of parameters could be selected for investigation. An approach using orthogonal array design of experiment was used to maximise the number of features to be investigated. The factors were ranked for ease of varying in the experiment. A CFD method was adopted for its suitability to test small changes in port form. Cast samples of ports would have too much variation that could mask the effect of changing factors.

The parameters to test are shown in Table 2 and some are shown in Figure 3.

Table 2 Factors for the port experimentation

Port Position in Orthogonal Axes. (X,Y,Z)

 Port position relative to major engine axis, in all orthogonal directions: X (the longitudinal axis); Y (across the engine) and Z (the vertical axis). The position of the port has been considered to be important in previous studies.

Position of Lip Surface. (W)

The offset of this surface is considered important to swirl and is shown in figure 3 (a).

Valve Seat Blend. (S_b)

Benchmarking this port feature showed that valve seat to throat blend varied with manufacturers. The different approaches are shown in Figure 3 (b).

Lip Radius. (R_l)

The breakaway of the air stream from the lip corner was deemed to be an important factor in the performance of the port. The radius can be seen in Figure 3(a)

Port Roughness. (R_g)

Roughness in the lip area was varied to investigate the effects of flow separation.

Valve Seat Alignment. (S_a)

The alignment of machined features of the port to cast features is a function of the machining facility. It was included to investigate the tight tolerances on machining whilst the rest of the port is subject to relatively large casting tolerances. It is shown in Figure 3 (c).

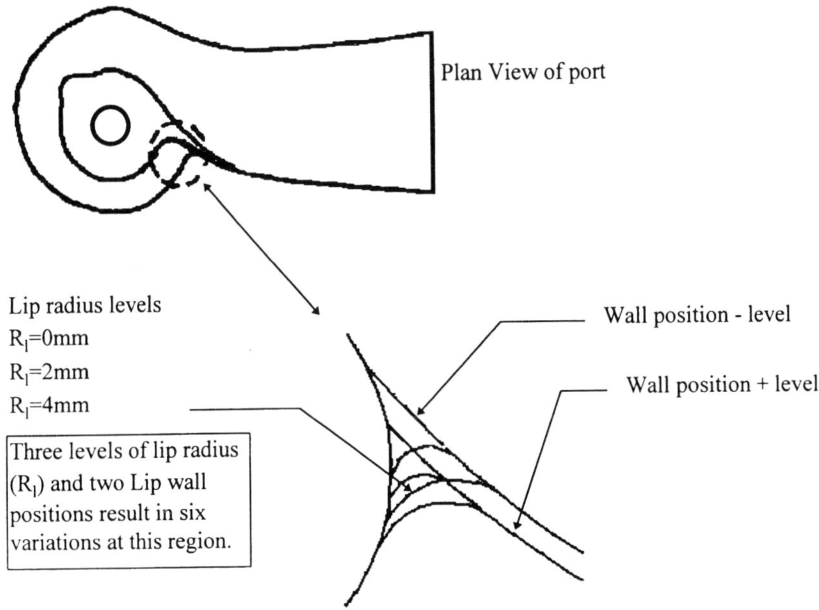

a) Port Lip wall and radius combinations

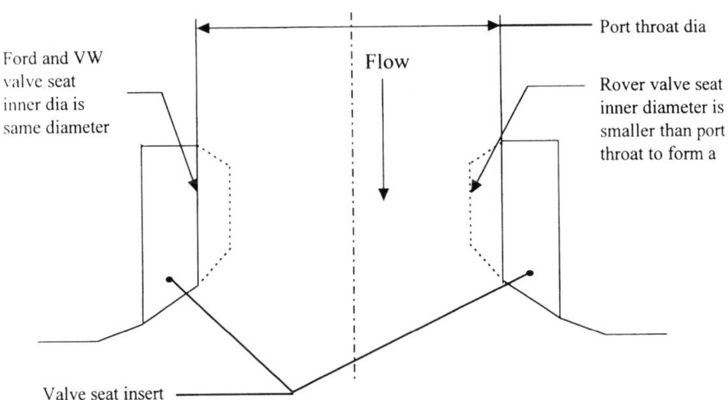

b) Valve seat blend

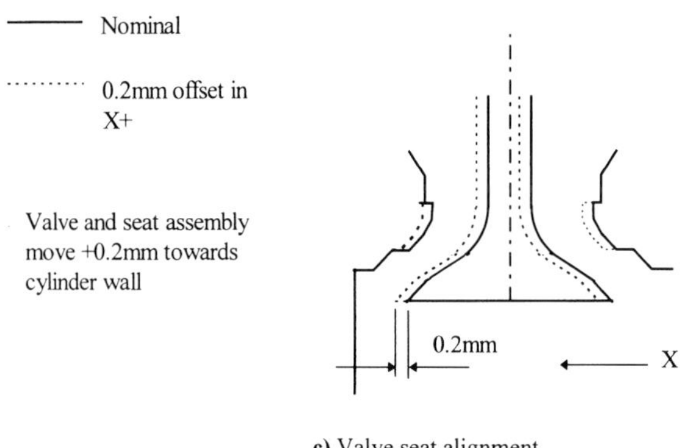

c) Valve seat alignment

Fig. 3 Factors for the CFD experimentation

The baseline model was based on a two valve engine, in which the single inlet port had a similar form to the ports in the twin port head. It was constructed and developed using nominal design of all components; port, valve, valve seat and cylinder head surface. The vane anemometer rig was modelled for recreating its flow and sensing conditions. Modelling was carried out using Star CD (4) analysis and PROSTAR(5) for meshing. The region near the valve seat was refined to minimise the inadequacy of the k–ε modelling in areas of high adverse pressure gradients as reported by Chen et al. (6) and Naser and Gossman (7). An added advantage was flexibility in modelling small changes for the study of variation of the features. The model contains 926,000 elements, of which 152,000 are included in the refinements near the valve seat.

CFD runs are relatively expensive, for complex models, and multiple runs for different valve lifts for each experimental model add to high cost. To resolve this, a single valve lift was determined for testing by using a signal-to-noise (S/N) ratio approach. The swirl readings, from a physical flow rig, for a variety of valve lifts were normalised to the ports mean swirl (signal) value. Then the spread (or noise), based on standard deviation, of the swirl was considered and used to determine a signal-noise ratio. The most representative reading occurred at a valve lift of 7mm, as shown in Figure 4. This was the valve lift adopted for the CFD model.

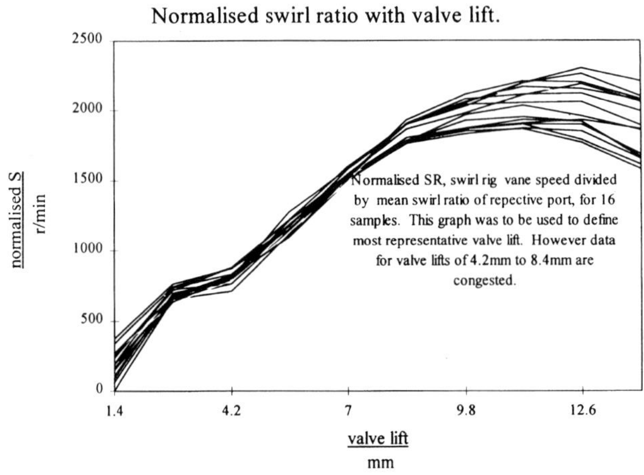

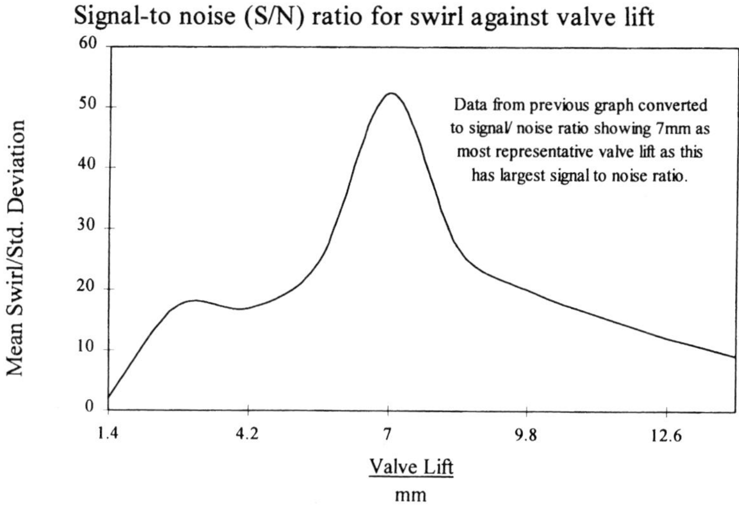

Fig. 4 Determination of single valve lift for CFD experiment

2.3 Designing the experimental array

The port lip radius and port position in the vertical, Z, axis, at three levels, and six features, at two levels, results in 576 (3^2 x 2^6) possible combinations for testing. The number of runs required was therefore reduced using orthogonal array testing. This method allows changing more than one factor at a time during testing but controlled such that the resultant test array maintains a balance between factors tested. The results from the array can be processed to extract information on individual factor effects.

The testing array is primarily a two level factor experiment, based on an orthogonal array design (8) with some factors combined, and collapsed, to give three factor tests. This combination of factors needs a minimum size array of 16 runs, termed a L_{16} array, to allow eight factors to be tested with two factors at three levels. A column combining and collapsing technique was adopted from Grove and Davis (9) to create three level factors from two level arrays and this reduced the number of columns in the array from fifteen to eleven, eight with factors and three empty columns. The empty columns should yield a null response and serve to set the baseline error in the test methodology.

The total number of experimental of the experimental runs is only a small fraction, $^1/_{36}$, of total combinations. Such a small fraction indicates the main effects will be confounded with interactions and are not resolved in detail for this screening experiment. The port lip radius, at three levels, and the port lip wall position, at two levels, give six combination levels for the port model in the lip area and are shown in Figure 3(a). The experimental run array is shown in the next section.

The runs were conducted to maximise ease of modelling changes in the CFD settings. Randomisation of the runs was not considered necessary for a computer based model.

Computational time for a CFD model was from two to three weeks. Therefore parallel processing on different computers was actioned though the alternative of using a mainframe was considered. (The computers used were IBM RS6000/590 workstations using 370Mb of core memory.) This was a benefit of computer modelling as reproducibility of computers running the same program could be guaranteed. If rig testing had been adopted this would not have been possible.

3. RESULTS

3.1 Baseline CFD model

Results from the baseline CFD runs were checked against those from the five samples tested on the anemometer rig as shown in Table 3.

Table 3 Comparison of CFD and flow rig results for baseline port design

	Swirl Ratio	% difference from measured value.	Flow coefficient	% difference from measured value.
Sample mean Rig measurement	2.16	0	.454	0
Sample range Rig measurement	2.11 -2.22		.45-.458	
CFD calculation (smooth surface)	2.21	+2..3%	.489	+7.7%
CFD calculation (cast iron simulation.*)	2.12	-1.9%	.472	+4.0%

*Cast iron surface represented by .25 sand grain roughness.

It is clear from Table 3 that the CFD predictions are in very good agreement with swirl and reasonable agreement with flow coefficient. This is particularly true for the present study in which the main aim was to investigate important features for sensitivity not to determine absolute values.

3.2 CFD results

Output from the CFD runs is in terms of the values of the swirl ratio and equivalent flow area (as a measure of flow efficiency). Other important parameters, such as gas velocity, direction, pressure and turbulent kinetic energy can be obtained by post processing of the flow field. This is an advantage of the CFD method over rig flow tests where only averaged swirl and flow performance values would have been obtained.

The test array, with description of the factors, with the CFD predicted results of swirl ratio and flow coefficient results is presented in Table 4.

Table 4 CFD test array and results

Run	Lip radius (R)	Local roughness (Rg)	Lip wall position (W)	Port position X axis (X)	Port position Y axis (Y)	Port position Z axis (Z)	Valve seat alignment (Sa)	Valve seat blend (Sb)	empty column (e1)	empty column (e2)	empty column (e3)	Swirl Ratio	Flow coefficient
1	-	+	+	-	+	-	+	-	-	-	+	2.89	0.4220
2	0	-	+	+	-	-	+	+	+	-	-	2.32	0.4379
3	0	+	-	+	+	-	-	+	-	+	-	2.07	0.4731
4	+	-	-	-	-	-	-	-	+	+	+	1.67	0.4935
5	-	-	-	+	+	0	+	-	+	+	-	2.42	0.4402
6	0	+	-	-	-	0	+	+	-	+	+	2.16	0.4765
7	0	-	+	-	+	0	-	+	+	-	+	2.27	0.4413
8	+	+	+	+	-	0	-	-	-	-	-	1.84	0.4731
9	-	+	+	-	-	0	-	+	+	+	-	3.01	0.4118
10	0	-	+	+	+	0	-	-	-	+	+	2.28	0.4447
11	0	+	-	+	-	0	+	-	+	-	+	2.09	0.4788
12	+	-	-	-	+	0	+	+	-	-	-	1.75	0.5015
13	-	-	-	+	-	+	-	+	-	-	+	2.47	0.4334
14	0	+	-	-	+	+	-	-	+	-	-	2.04	0.4742
15	0	-	+	-	-	+	+	-	-	+	-	2.30	0.4379
16	+	+	+	+	+	+	+	+	+	+	+	1.83	0.4742

The reduction in turbulent kinetic energy down the cylinder bore for run 1 is shown in figure 5. It can be seen the flow becomes mainly uniform at the vane position at z =-147mm. All other 15 runs show a similar trend. This gives confidence that the computer model vane position is in a region of solid body rotation, which supports one of the assumptions of swirl ratio measurement.

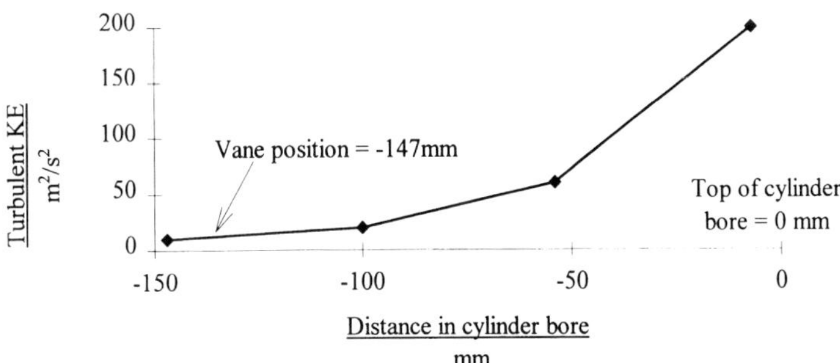

Fig. 5 Reduction of turbulent kinetic energy down cylinder bore

Effect plots of factors showing their influence on swirl and flow coefficient can be seen in Figure 6.

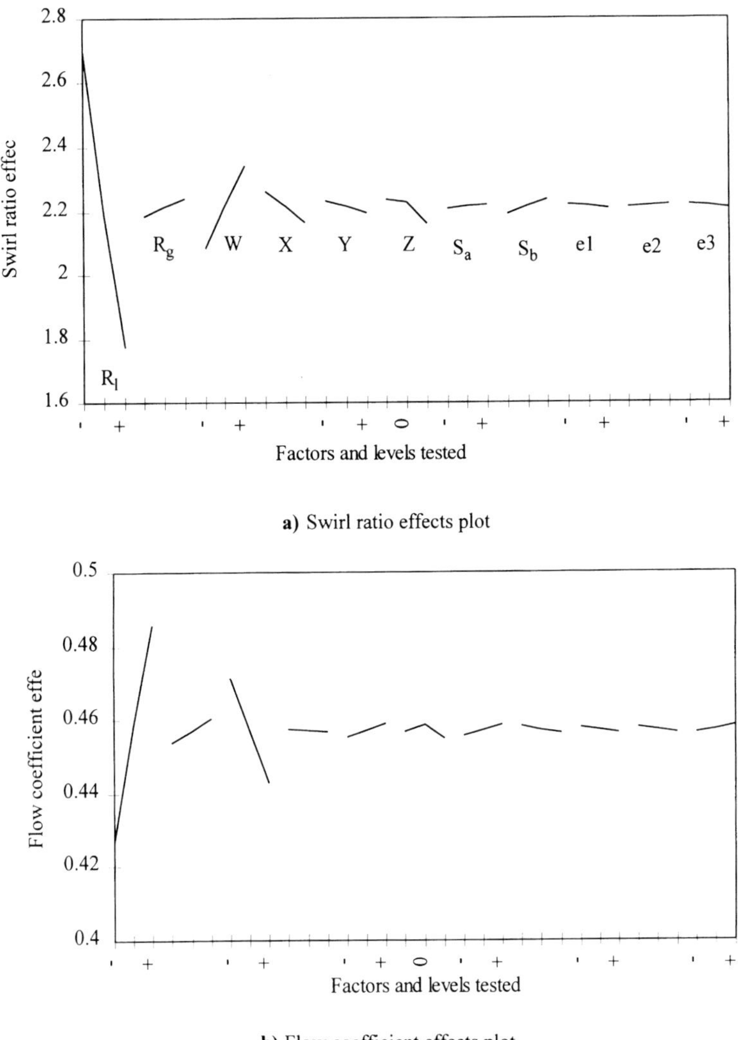

a) Swirl ratio effects plot

b) Flow coefficient effects plot

Fig. 6 Effects plots for swirl ratio and flow coefficient

S608/005 © IMechE 1999

Evaluation of the CFD plots showed the lip radius and lip wall position having a major impact on the flow quality in the swirl chamber section above the valve throat. This can also be seen from the velocity vector plots shown in Figure 7, the amount of stagnant flow, adjacent to the valve stem, is proportional to the lip radius. Comparisons of flow diagrams for the intermediary swirl results show a proportional amount of stagnant flow at the valve stem.

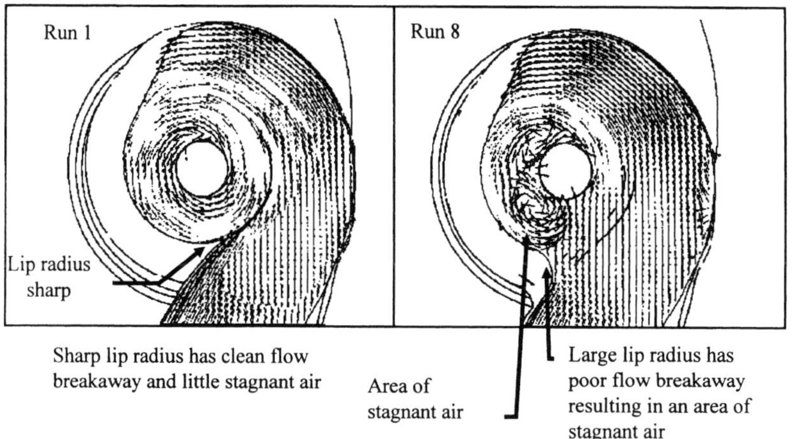

Sharp lip radius has clean flow breakaway and little stagnant air

Area of stagnant air

Large lip radius has poor flow breakaway resulting in an area of stagnant air

Fig.7 Lip radius influence on swirl motion

3.3 Linear regression analysis
Initially, all factors were considered, even the empty columns, as they are a combination of different interactions, albeit complex. The t-test was used as a check for factor significance. A second regression was conducted only considering the factors that showed significance in the t-test values from the regression using all factors.

For swirl ratio the analysis showed lip radius and wall position as significant factors in a t-test. The resultant prediction equation for swirl ratio, using coded values of lip radius and wall position from the experiment, is;

$$\text{Swirl Ratio} = 2.213 - 0.4625R_1 + 0.1294\,W \tag{1}$$

For flow coefficient, lip wall position and lip radius had the largest effect on the flow. Local roughness (R_g) was also significant but to a much smaller extent with a t-test value of 3.46 slightly above the critical value of 2.78. The resultant flow coefficient equation using the coefficients for significant effects only is;

$$\text{Flow coefficient} = 0.45715 + 0.02936R_1 - 0.0\,1425W + 0.00333Rg \tag{2}$$

The vast amount of data held in a CFD model of nearly one million cells has to be reduced to manageable quantitative data for analysis in a DOE. In this study only swirl ratio and flow coefficient were chosen and presented, but all data is held in the CFD output files which can be used in further studies.

4. DISCUSSION

4.1 Discussion of the results

From CFD results shown in Figure 8, a linear relationship between swirl and flow coefficient gives the best correlation coefficient, R^2, of 0.8494, over the range of swirl and flow tested in the study. This supports the theory that some of the potential energy is converted to kinetic energy in the swirling air. However, Gale (10) reports a quadratic relationship between swirl and flow coefficient over a wider range when various ports were benchmarked.

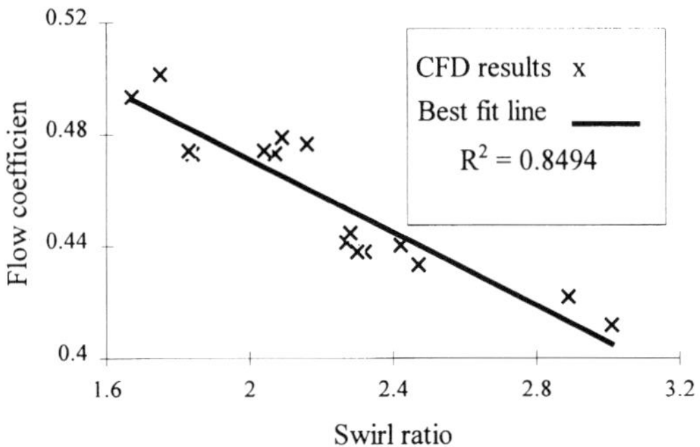

Fig. 8 Swirl and flow area for CFD experiment

It can be seen, in Appendix 1, flow coefficient is inversely proportional to swirl ratio. It is therefore not surprising the closeness of the individual models to the general line of best fit. The variation of flow efficiency for a given swirl is approximately 3%.

It should be recognised direct comparisons with published data on other port investigations are of limited usefulness as test rig details and calculation assumptions are not standardised.

4.2 Effects of factors

Swirl Ratio Significant Effect Factors

Effects plots and regression analysis of the data show lip wall position and lip radius are highly significant compared to the other factors investigated. The position of the port in the three axes is not critical over the range tested. This was initially surprising but may be explained by considering the working mechanism of the port. The helical section is a swirl forming chamber. Bodily movement of this will not prevent the swirl from forming. It will only affect the exit of the swirling air through the valve seat. The port movement in the horizontal directions X and Y will effectively change the angle of the port relative to the cylinder centreline, but small changes in position will have only a small effect compared to the relatively large dimension of the port. The valve seat appears to be acting as a controlling influence on air flow direction into the cylinder. The lip angle position partly affects position in the X direction. As this local area moves it has a small effect on the mean position of the port but this is ignored in the present analysis. However it is deemed that neglecting this should not have a large effect on the predicted swirl and flow coefficient since the X axis has been shown to be an insignificant factor. The lip position, rather than the port position, appears to be the major factor in the generation of non productive flow adjacent to the valve stem.

Flow Coefficient Significant Effect Factors
The results were further analysed to see if any factors could improve flow whilst maintaining swirl ratio. The only feature identified was roughness of the area adjacent to the lip. The roughness in the CFD model is not directly equivalent to surface texture roughness of the production port and more investigation into the form of a feature to break up the boundary layer that would not carbon up in service is required.

This DOE was primarily a tolerance study on features of the port. Therefore the lip radius and lip wall position are the important tolerances to control for reducing variation in port performance. A further improvement is to eliminate the variation of tolerances. For the lip radius, a method would be to machine a sharp corner at the lip, this is equivalent to R_1 at minus level of CFD experimentation. This would make the radius term, R_1, in equations 1 and 2 constants, thus reducing the total variation of swirl and flow efficiency. Reduction of variation by this method is subject to a patent application (11).

The drive for improved quality of port performance must be focused more earlier in the design process. This study investigated variation on an existing prototype port. Some features identified as worthy of investigation had to be omitted from the experimental plan because of the difficulty of changing the model. Widener (12) reports techniques for reducing the number of parameters defining inlet ports from typically 600 to 150 that will assist true parametric design studies to overcome this limitation. In addition to this the increases in computing power will allow more complex CFD studies to be conducted.

## 5.	CONCLUSIONS

The following conclusions can be drawn from this study;
- The lip radius and the lip wall position have been found to have the most influence on the performance of the port and the port performance variation can be reduced by controlling these features.
- The CFD analysis of the port has benefits of ability to assess port performance prior to part manufacture and to output information not easily obtainable from rig testing.
- Orthogonal arrays can be used in a design of experiment that vastly increases effectiveness of testing.

ACKNOWLEDGEMENTS

The authors acknowledge the contributions made by all colleagues at Ford Motor Company and the University of Bradford who made this study possible. Acknowledgement is also due to the consultants Adapco for their support in conducting the CFD analysis.

REFERENCES

1   **Needham, J.R.** and **Whelan, S.,** Meeting the Challenge of Low Emissions and Fuel Economy with the Ricardo Four Valve High Speed Direct Injection Engine., *Proc. Instn Mech. Engrs,* Transactions, Part D, 1994 **208,** 181 -190

2   **Herrmann H.-O. and Dürnholz, M.,** Development of a DI-Diesel Engine with Four Valves for Passenger Cars, *SAE paper,* 1995, 950808, 109-119

3   **Stone, C.R.** and **Ladommatos, N.,** The Measurement and Analysis of Swirl in Steady State Flow., *SAE paper,* 1992, 921642

4   **Star CD,** Version 2.300, Computational Dynamics Ltd, London

5   **PROSTAR,** Computational Dynamics Ltd, London

6   **Chen, A., Ganti, G., Lee, K.C.** and **Vianneskis, M.,** A comparison of CFD Predictions and LDA Measurements of the Flow Through a Generic Inlet Port., *American Society of Mechanical Engineers (ASME,),* 1994,Vol 8- Part A, 85-93.

7   **Nasser, J.A.** and **Gosman, A.D.,** Flow prediction in an axisymmetric inlet valve/port assembly using variants of k-ε., *Proc. Instn Mech. Engrs* 1995, 209, Part D, 69.

8   **Buswell, N.,** User Guide to 2-Level Orthogaonal Arrays, *Ford Motor Co.* (Internal Document), 1994.

9   **Grove, D.M.** and **Davis, T.,** *Engineering Quality and Experimental Design,* 1992 (Longman Scientific and Technical, London)

10   **Gale, N.F.,** Diesel Engine Cylinder Head Design: The compromises and the techniques. *SAE paper,* 1990, 900133, 1-24.

11   **Brignall, A., Turner, P.N.** and **Baker, P.,** Inlet Port with Sharp Edge Swirl Lip. Patent Application GB9727175.3

12   **Widener, S.K.,** Parametric Design of Helical Intake Ports. *SAE paper,* 1995, 950818.

The importance of manufacturing tolerances on inlet port performance

B M TRAIN
Ricardo Consulting Engineers Limited, Shoreham-by-Sea, UK

ABSTRACT

A statistically based methodology for assessing the effect of manufacturing tolerances on inlet port performance has been developed. Tolerance extremes are assessed on a steady-state flow rig to determine the flow, swirl/tumble characteristics. Any improvements to production processes to control port performance and their benefit can be quantified. Such techniques are particularly valuable to combustion systems which have high specific air motion requirement, such as in swirl based G-DI engines.

1.   INTRODUCTION

Multi-valve engines began to take an increasing market share from the early 1980's onward. As the demands on the combustion systems grew, inlet port configurations were developed to generate tumbling (or barrel swirl) air motion, Reference 1. Some researchers developed ports which produced swirling air motion by means of port deactivation, Reference 2. These systems are relatively complex and costly to manufacture.

Recently there has been much activity in developing gasoline direct injection (G-DI) combustion systems. Mitsubishi where first into the market place using a fixed geometry, reverse tumble system, Reference 3. Ricardo have demonstrated that this type of approach, Figure 1, offers minimum incremental manufacturing cost and complexity over a conventional fixed geometry four valve/cylinder engine whilst giving very good stratified charge performance, References 4 and 5. However, the majority of alternative systems rely on variable air motion by means of variable swirl control valves, as typified by the Toyota system, Figure 2, Reference 6. Our own research has shown that a swirling system requires good control of swirl level to provide good combustion over a range of speeds and loads under stratified charge operation, Figure 3.

In a multi-cylinder engine it is clearly very important to minimise the cylinder to cylinder variation in air motion. Whilst this is obviously influenced by assembly of the swirl control valves, the port manufacturing tolerances can also have a significant effect. The latter is true of all ports, particularly those which generate significant air motion, but in the case of a swirling stratified charge engine swirl variations can make the difference between combustion and no combustion. Clearly, it is important to establish the variation which could be expected in production and if required be able to investigate the benefits of improved production processes as early as possible in the development stages. We have, over the last six years, developed a statistical methodology for this purpose. The methodology is equally applicable to tumbling engines as well as swirling engines.

2   DESIGN OF EXPERIMENT

Manufacturing tolerances, casting and machining, mean production cylinder heads will always vary to some degree from the nominal design intent. The approach developed for assessing the influences of these variations typically involves the following steps:

- Selection of variables
- Design of Experiment
- Manufacture of flow boxes
- Testing of flowboxes
- Analysis of Results
- Confirmation test
- Conclusions and Recommendations

2.1   Selection of variables

There are many potential variables in the manufacture of an inlet port. However, to investigate all of these would be a large and costly exercise, even using a design of experiment (DoE) approach. For a successful design of experiment variables have to be chosen based on the following criteria:

- Each variable should be expected to have a significant effect on the air motion and/or flow characteristics
- No one parameter should be so significant as to hide the effect of the others
- The influence within the tolerance range will be substantially linear

It is possible to group together some individual parameters, particularly if the production processes will tend to affect them in a similar way.

During the development of a particular port, information will have been obtained relating to the sensitivity of various features. Other simple one off experiments can also be carried out to help this understanding or if no prior knowledge of the port exists Typical variables include:

- Port location
- Chamber location
- Port throat cutter depth
- Valve seat machining
- Port throat diameter
- Cylinder bore offset
- Exhaust valve head protrusion
- Port location at manifold face

                    S608/006 © IMechE 1999

It is also necessary to have an understanding of the direction in which the tolerance should move in order to have a positive or negative effect. To take a simple case as an example, if the chamber is quite open with the inlet valves close to the bore; bore offset will have an effect in X and Y directions, but obviously not Z. The direction to most enhance swirl (and reduce flow) would be a combined movement in X and Y as shown, Figure 4.

2.2   Design of Experiment

There are a number of different statistical design of experiments to chose from. For the purposes of investigating port tolerance effects Ricardo have found that a Taguchi style Design of Experiment (DoE) is preferred. This is because the relative influence of any tolerance does not have to be known in a advance and the number of experiments (and hence flowboxes) can be controlled. A typical array involves the use of eight flowboxes, such as that used in Case A, Table 1. In this case the experiment enables all the major interactions to be resolved within the main experiment. The matrix is a hybrid design incorporating a five factor experiment in 16 tests and a sub-experiment on bore offset. Although there are twenty tests to be performed, only eight flowboxes were required. The machined mask clearance being initially tested in the maximum material condition and then remachined to the minimum condition. Bore offset is easily adjusted without changing the flowbox. In such a case, the derived relationships would be confirmed by producing two flowboxes at previously untested combinations of tolerance extremes.

The experiment involves assessment of flowboxes manufactured to represent different combinations of tolerance extremes as shown by the plus and minus values in the matrix. Statistical or arithmetic tolerances can be used although, experimentally, the larger arithmetic tolerances are preferred to maximise the differences in swirl and flow and hence the accuracy of the experiment.

Table 1. Design of Experiment Matrix - Case A

Flowbox	Test	Valve seat to Chamber in Y	Gas face toChamber in Z	Valve seat to Chamber in X	Throat and seat envelope	Machined mask clearance	Bore Offset
1	1	-1	1	-1	1	-1	-1
1	2	-1	1	-1	1	-1	1
1	3	-1	1	-1	1	1	-1
1	4	-1	1	-1	1	1	1
2	5	-1	1	1	-1	-1	-1
2	6	-1	1	1	-1	-1	1
2	7	-1	1	1	-1	1	-1
2	8	-1	1	1	-1	1	1
3	9	-1	-1	-1	-1	-1	-1
3	10	-1	-1	-1	-1	1	-1
4	11	-1	-1	1	1	-1	-1
4	12	-1	-1	1	1	1	-1
5	13	1	-1	-1	1	-1	-1
5	14	1	-1	-1	1	1	-1
6	15	1	-1	1	-1	-1	-1
6	16	1	-1	1	-1	1	-1
7	17	1	1	-1	-1	-1	-1
7	18	1	1	-1	-1	1	-1
8	19	1	1	1	1	-1	-1
8	20	1	1	1	1	1	-1

A more unusual case is illustrated in Table 2, whereby there where two configurations of swirl port within a cylinder head. In this case, a full factorial experiment could not be carried out with the possibility of some interactions requiring subsequent resolution. The tolerances affecting a given variable are combined to give a maximum possible out of position. The tolerances applicable to Case B are given in Table 3.

Table 2. Design of Experiment Matrix - Case B

Flow box	Hardware Variables				Software
	1	2	3	4	5
1	-1	-1	-1	-1	-1
	-1	-1	-1	-1	1
2	-1	-1	1	1	-1
	-1	-1	1	1	1
3	-1	1	-1	1	-1
	-1	1	-1	1	1
4	-1	1	1	-1	-1
	-1	1	1	-1	1
5	1	-1	-1	1	-1
	1	-1	-1	1	1
6	1	-1	1	-1	-1
	1	-1	1	-1	1
7	1	1	-1	-1	-1
	1	1	-1	-1	1
8	1	1	1	1	-1
	1	1	1	1	1

Table 3. Combined Offsets at Tolerance Extremes

Variable No.	Variable Name	Level +1			Level -1		
		X	Y	Z	X	Y	Z
1	Port type	Port 1,3			Port 2,4		
2	Port XY	0.792	0.792	0	-0.792	-0.792	0
3	Port Z	0	0	1.24	0	0	-0.21
4	Chamber XYZ	0.37	0	0	-0.37	0	0.3
5	Bore offset XY	0.15	0.15	0	-0.15	0.15*	0

The DoE in Figure 2 is an unrandomised split plot design. This leaves possible 2 way interactions which can be resolved by a further four box confirmation experiment. In this case the main DoE would derive initial polynomial expressions for swirl ratio and gulp factor, hence a D-optimal DoE, which requires some prior knowledge of the answer, could be used involving four flowboxes, Table 4. This approach was selected as it could most effectively discriminate the interaction effects.

Table 4. D-Optimal Confirmation DoE Matrix

Flow box	Hardware Variables				Software	Swirl Ratio	Gulp Factor
	1	2	3	4	5		
9	-1	-1	1	-1	-1	1.16	0.605
10	-1	1	-1	-1	-1	1.35	0.641
11	1	-1	-1	-1	-1	1.2	0.599
12	1	1	1	-1	-1	1.51	0.645

2.3 Manufacture and Testing of Flowboxes

The flowboxes are manufactured using CNC machine tools. Prior to this CAD models are created with the desired offsets in place. During this process it is sometimes found that some theoretical tolerances are impossible to achieve in practise and so some revision may be required. The machining of the boxes must be carefully controlled to ensure that any tolerances in their manufacture are insignificant compared to the tolerances being studied. A typical example of inspection data from a DoE flowbox is given in Table 5. The table lists the offsets at the tolerance extremes for that particular box and shows the variation from the setting achieved during the manufacture of the flowbox. In all cases the errors were small compared to the required setting.

Table 5. Typical Flowbox Inspection Data

Parameter	Set Location Rel. to Nominal (mm)	Measured Error (mm)	Error (%)
Chamber in X	1.16	0.04	3.44
Chamber in Y	0.79	0.03	3.79
Chamber in Z	0.51	0.01	1.96
Seat In X	0.70	0.07	10.00
Seat in Y	0.79	0.03	3.79
Seat in Z	0.43	0.00	0.00

The flowboxes are tested on a steady state air flow rig shown in Figure 5. This type of rig is quick and relatively cost effective to use. Accuracy is maintained by rigorous quality control procedures and regular benchmarking of a 'standard' cylinder head. The differences in air motion about the nominal for this type of experiment are meaningful and have been shown to correspond to difference in combustion and power. However, for a complete assessment of new concepts the use of a dynamic flow rig, Reference 7, is recommended. The boxes are tested in a random order and subject to repeat testing. The characteristics used to define the port performance are swirl ratio and gulp factor. An increase in gulp factor represents a decrease in flow capacity.

Figure 6 contains the air flow rig test results from Case B. This clearly illustrates the tendency for high swirl to result in low flow (high gulp factor). There is large scatter but no combinations result in high swirl and high flow or vice versa.

2.4   Analysis of Results

Analysis of the results from the DoE produces a polynomial expression for the flow and swirl characteristics in terms of the tolerance extremes such as shown in the simplified, two variable, example below:

- Swirl Ratio = A+ B*Variable 1 tolerance + C* Variable 2 tolerance
- Gulp Factor = D+ E* Variable 1 tolerance + F* Variable 2 tolerance

In this case if A,B,C,D,E all equal 1 and all the tolerances are in the +1 condition, a flowbox made to this configuration would have a swirl ratio and gulp factor of 1+1+1 = 3.

The expressions derived from the total experiment for Case B are:

- Swirl ratio = 1.303 - 0.0277*Port type + 0.2019*Port XY + 0.0225*Port Z -
- 0.0394*Port type*Port Z + 0.0410*Port XY*Port Z + 0.0375*Port XY*Chamber XYZ

- Gulp factor = 0.6276 + 0.0013*Port type + 0.0194*Port XY - 0.0028*Chamber XY -
  0.0021*Port XY*Port Z - 0.0025Port XY*Chamber XYZ - 0.004*Port Z*Chamber XYZ

This indicates that the most important variable to control both swirl and gulp factor is the location of the port in X and Y. In order to validate the expression the swirl and gulp factors were calculated for the nominal conditions, for which flowboxes already existed. The results were:

	Port A	Port B
Predicted Swirl Ratio	1.28	1.33
Measured Swirl Ratio	1.32	1.36
Predicted Gulp Factor	0.629	0.626
Measured Gulp Factor	0.622	0.621

The resultant relationship arising from the experimental design was able to predict the results from the master flow boxes (nominal locations). The maximum errors were approximately 1% on gulp factor and 3% on swirl ratio.

A batch of prototype heads were procured and the heads with the largest visual casting errors picked out for detail inspection. The flow and gulp factors are compared with the flow boxes in Figure 7. This demonstrates that the heads were within the total predicted spread. The inspection data allowed the swirl ratio to be calculated using the DoE expression. The measured and predicted results, cross-plotted in Figure 8, shows good agreement.

2.5   Reduction of Variation

The expression derived from the DoE can be used to assess the acceptability of the expected production variation. If this is not acceptable various scenarios for improving the manufacturing process can be examined and their benefits quantified. In Case B the tolerances were examined and areas for improvement identified. In this case the largest improvement could be obtained in the casting process. Figure 9 compares the original flow box data with those predicted with improved tolerances. This shows that the predicted effect would be:

- Reduction in swirl ratio spread from 0.54 Rs to 0.1 Rs
- Reduction in Gulp factor spread from 0.052 to 0.014

3 CONCLUSIONS

Manufacturing tolerances can result in large variations in inlet port performance to the detriment of engine operation.

A design of experiment methodology has been developed to enable the production variation to be predicted with a high degree of confidence for both tumbling and swirling air motion.

The resultant expressions arising from the DoE can be used to investigate possible improvements to the production process.

4 REFERENCES

1. C. D. de Boer, R.J.R. Johns, D.W, Grigg, B.M. Train, I. Denbratt, J-R. Linna, 'Refinement with Performance and Economy for Four Valve Automotive Engines', I.Mech.E C394/053

2. 'Ford 3.0 Liter Duratec V6, Automotive Engineering, Jan 1996

3. Y. Iwamoto, K. Noma, O. Nakayama, T. Yamauchi, H. Ando ' Development of Direct Injection Gasoline Engine', SAE 970541

4. N.S. Jackson, J. Stokes, P.A. Whitaker, T.H. Lake, 'A Direct Injection Stratified Charge Gasoline Combustion System for Future European Passenger Cars' I.Mech.E Lean Burn Combustion Engines Seminar 3-4 Dec 1996

5. T.H. Lake, J. Stokes, P.A. Whitaker, J.V. Crump, 'Comparison of Direct Injection Combustion Systems', SAE 980154

6. J. Harada, T. Tomita, H. Mizuno, Z. Mashiki, Y. Ito, ' Development of Direct Injection Gasoline Engine', SAE 970540

7. N.S. Jackson, J. Stokes, M. Heikal, J. Downie, ' A Dynamic Water Flow Visualisation rig for Automotive Combustion System Development', SAE 950728

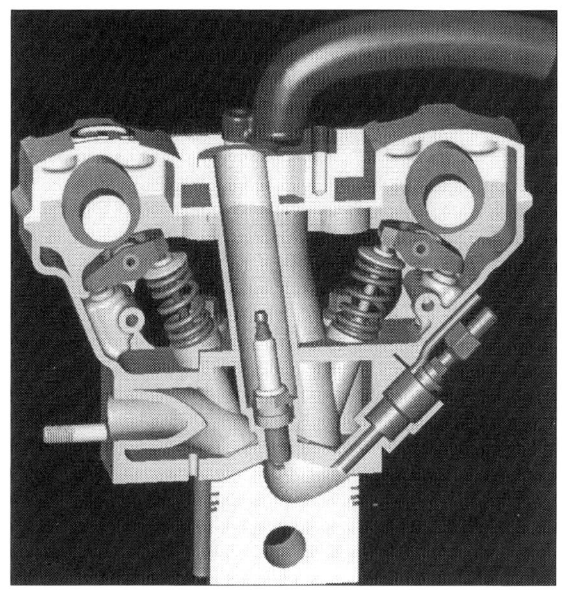

Figure 1. Top Entry G-DI

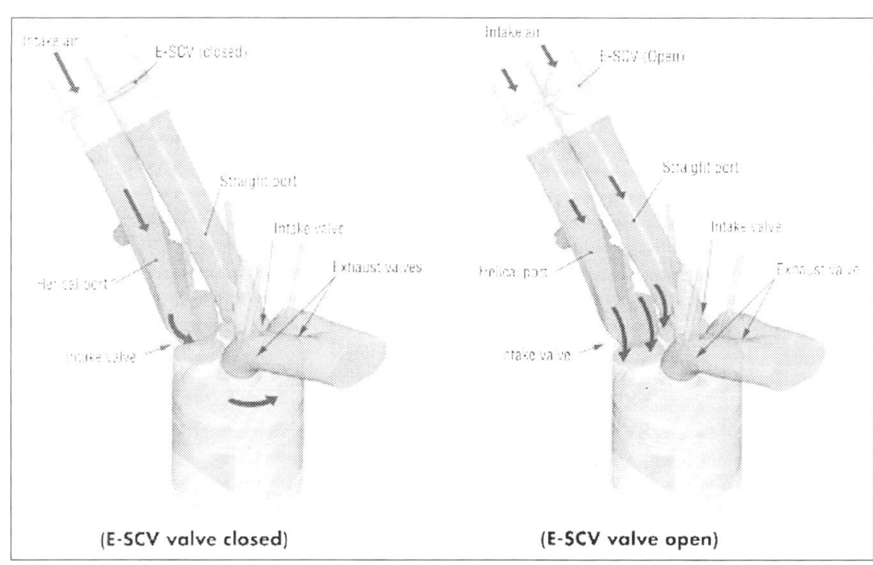

Source: Toyota

Figure 2 : Toyota D4 Swirl-type G-DI Inlet Ports with Swirl Control Valve (SCV)

  S608/006 © IMechE 1999

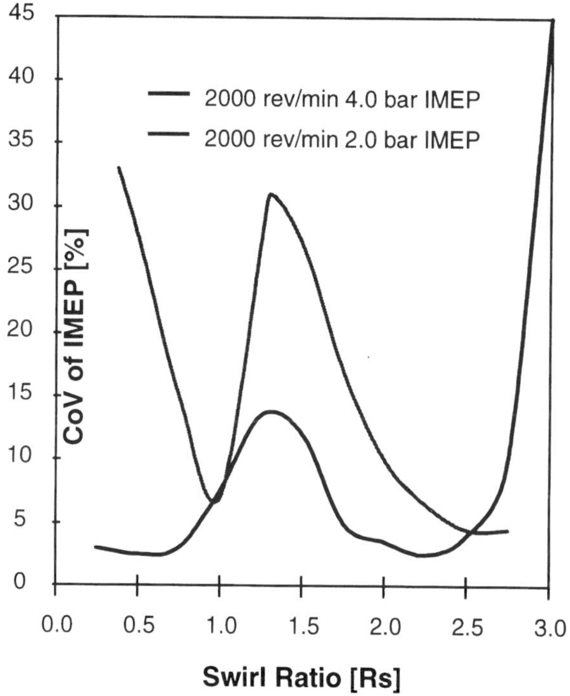

Figure 3. Typical G-DI Swirl Response Under Stratified Charge Operation

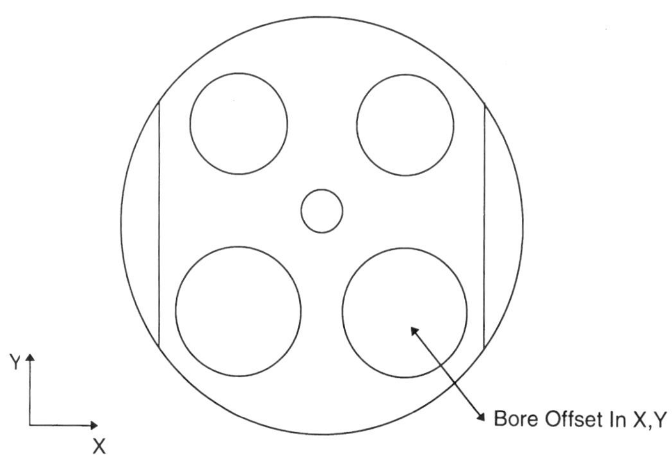

Figure 4 Bore Offset to Maximise Swirl Effect.

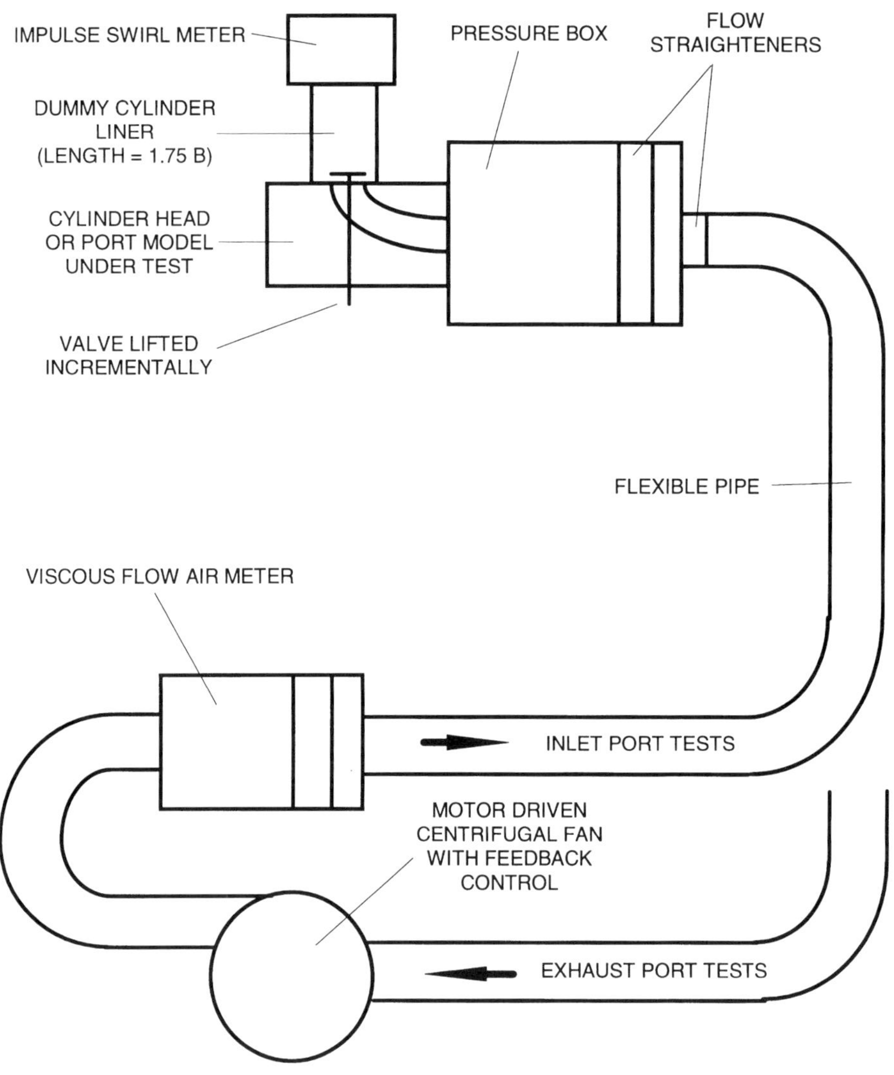

Figure 5. Steady state Air Flow Rig - Swirl Measurement

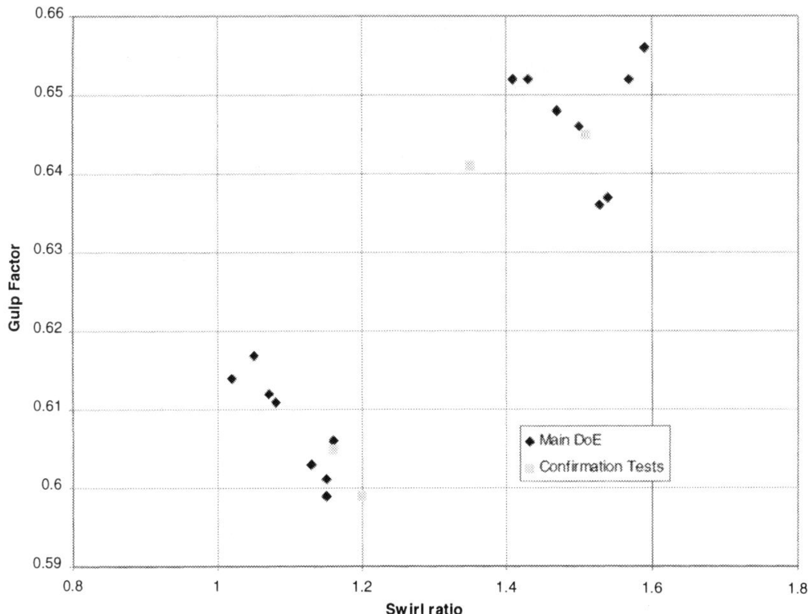

Figure 6. Measured variation in Swirl ratio and Gulp Factor

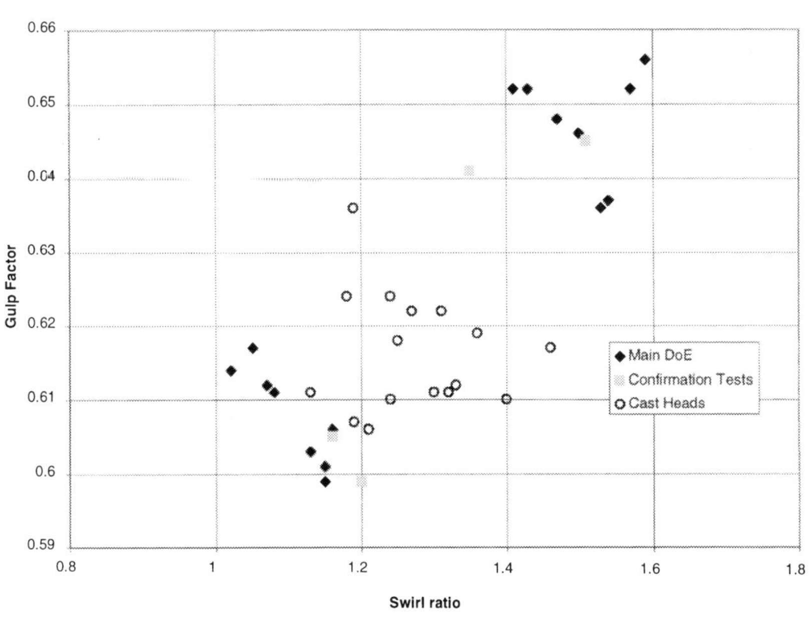

Figure 7. Comparison of Flowboxes and Prototype Cast Cylinder Heads

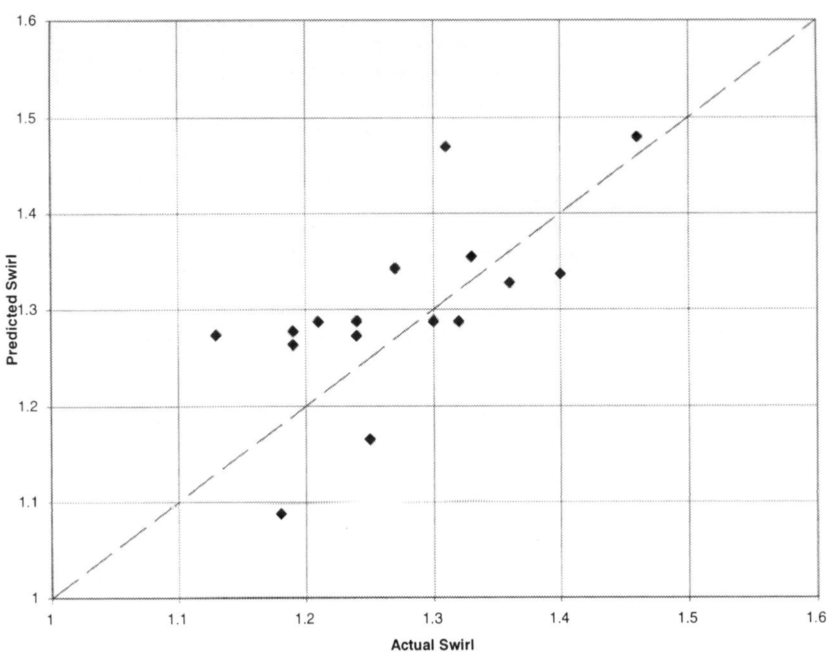

Figure 8. Measured and Predicted Swirl Ratio for the Prototype Cast Cylinder Heads

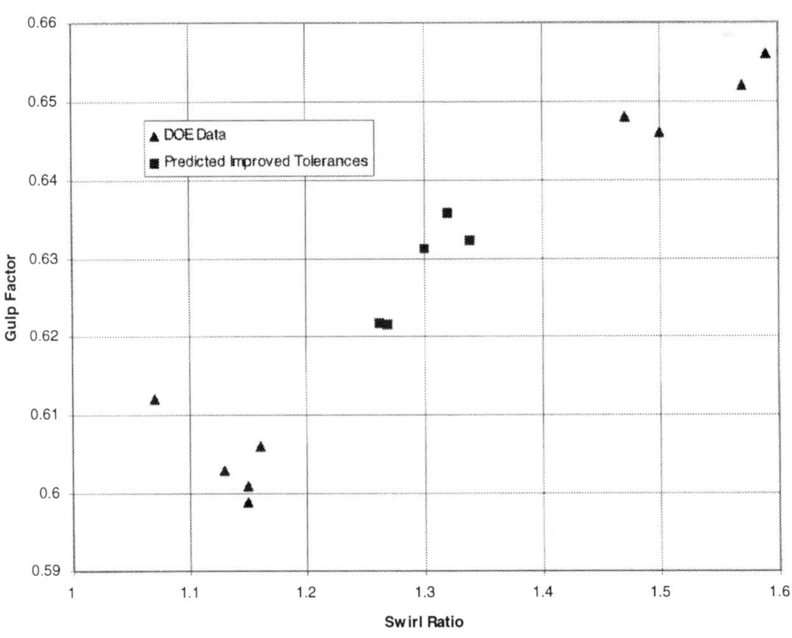

Figure 9. Effect of Reduced Tolerances on Flow and Swirl Variation (Predicted)

S608/006 © IMechE 1999

Effects of the variation of valve lift, duration, and phasing on the emissions and fuel economy of a multi-cylinder, four-valve SI engine

A GHAURI and **S H RICHARDSON**
Jaguar Cars Limited, Coventry, UK
C J E NIGHTINGALE
Department of Mechanical Engineering, University College London, UK

Introduction

Mechanisms that allow the control of camshaft phasing are becoming relatively commonplace on current designs of 4-valve spark ignition engines. These are employed to reduce the compromises that have to be made in camshaft design in order to have acceptable stability and emission levels at part load conditions combined with high volumetric efficiency throughout the speed range at full load. Camshaft phasing is more beneficial on 4-valve than 2-valve engines because the increased flow area gives higher volumetric efficiencies at full load, but making full use of this potential will lead to both loss of low-speed torque and also excessive residual gas concentration at part load, unless some change in phasing is applied. There are four possible strategies for camshaft phasing: intake camshaft only, exhaust camshaft only, intake and exhaust camshafts phased equally (Dual Equal), and intake and exhaust camshafts phased independently (Dual Independent). Leone et al investigated each strategy in turn and reported [1] on their relative advantages. An alternative approach was taken by Seabrook et al [2] in which the experiments were designed using statistics such that the effect of any variation in phasing could be predicted from the results of a relatively low number of selected test points.

An obvious development as designers strive to add further refinement to 4-valve engines is to introduce some method of altering camshaft lift and/or camshaft duration as well as, or instead of, phasing. This has already been achieved in production using a mechanism [3] that switches from a camshaft designed for part load to one with higher lift, increased duration, and different phasing for high speed and load operation. Ideally the settings would be continuously variable rather than having a two-position arrangement, and this has been conceived for duration alone through the adoption of an ingenious mechanism [4] which varies the speed of camshaft rotation. Wilson et al [5] have described the application of an electro-hydraulic valve actuation system which allows complete variation of valve events; their system provides an excellent tool for test bed development work but is impractical for production. Efforts have been continuing for years to develop electro-mechanical and electro-hydraulic systems for production, and may well succeed soon.

The work reported here describes an experimental investigation to evaluate the potential benefits in terms of reduced emissions, improved stability, and lower fuel consumption of being able to optimize phasing, lift, and duration for a range of low speed and load conditions. The work was performed on a modern design of 4-valve, Vee-8, engine and used the statistical approach to the design of the experiments which has been described in [2]. The method by which the changes to valve events would be achieved on a production engine was not a subject of this investigation, and so the changes were made simply by having a range of different camshafts. It was intended that the results would quantify the benefits of having a continuously-variable system as compared to a profile-switching arrangement. The investigation has been divided into two phases, the first covering symmetrical valve event strategies and the second asymmetric strategies (ie one inlet valve following a different camshaft profile to the other). It is the first phase that is reported here.

Engine and Instrumentation

The engine was an early prototype version of a production 4-litre, 90° Vee-8, unit with a pent-roof combustion chamber and 4 valves per cylinder. Some details of the engine are given in Table 1.

Intake camshaft timing (standard, retarded)	Valves open 5° aTDC
Exhaust camshaft timing (standard)	Valves open 45° bBDC
Intake camshaft phasing range	30°
Camshaft duration (standard)	240° (intake) 230° (exhaust)
Bore	86 mm
Stroke	86 mm
Compression ratio	10.75:1
Valves per cylinder	4
Fuel injection	Port

Table 1   Engine Specification

It was found more convenient to remove the variable phasing units on the intake camshafts for the testing and to set the timings of the intake and exhaust camshafts manually using a series of hardened setting blocks machined to an appropriate range of angles. The drive-by-wire throttle, normally under control of the engine management system, was replaced by a unit which enabled the tester to adjust the throttle angle through the action of a stepper motor. This, in turn, was replaced by a throttle unit which could be adjusted manually to obtain direct setting with the required precision for testing at idle. Certain other engine modifications were incorporated to facilitate the many camshaft changes that were required during the test programme. Provision was made to sample exhaust gases from each cylinder bank separately. Burn rates and cyclic variability were calculated from the pressure traces of a single cylinder. Some cylinder pressure traces were also recorded from the remaining cylinders of the same bank to check that the selected cylinder was representative.

As mentioned in the introductory section, variation of events was achieved by using a series of intake camshafts manufactured with differing durations and lifts. One common feature between the camshafts was the design strategy of dividing the lift curve into sectors, and working on the principle of scaling the acceleration characteristics for each of the sectors to obtain the appropriate cam design, with the effect of forming a 'family' of camshafts. This was adopted partly for convenience of design, and partly to avoid the introduction of any unforeseen variable resulting from a particular type of camshaft lift profile.

Design of Experiment
Selection of variables used
The following were identified as the engine responses of most interest: unburnt hydrocarbon (uHC) emissions, oxide of nitrogen (NOx) emissions, carbon monoxide (CO) emissions, fuel consumption, burn rate, cyclic variability and exhaust gas temperature. It was decided that the valve event variables of most interest were: intake camshaft phasing, intake camshaft duration, intake camshaft lift, and exhaust camshaft phasing. The minimum number of test points for each variable was identified as five, since it is likely that, at least, some of the variables would lead to non-linear responses. A full-factorial test matrix would require $5^4 = 625$ tests for each engine load and speed condition, clearly not practical in terms of camshaft manufacture and not even desirable due to likely difficulties in keeping a consistent engine performance over an extended test programme. Faced with a comparable situation, Wilson et al [5] used rig testing to identify eight promising camshaft design strategies worth testing on an engine. The danger of this approach is that some interesting effects may be overlooked as so few areas of the test matrix are visited. It was decided to use a Central Composite Rotatable Design (CCRD) of experiment, which enabled all the main effects and second-order interactions to be assessed for four variables at five levels using only 25 test points. The details of the valve event timings for each test point are given in Table 2. The procedure followed was to record the engine responses of interest from these 25 test points and to subject them to regression analysis in order to identify the significant variables and interactions. Equations were derived from this analysis so that the responses could be predicted with reasonable certainty at the remaining parts of the 625-point test matrix not visited during the 25 tests. The background to the application of statistics to engine experimentation, in particular the use of the CCRD, is described in more detail by Seabrook et al [2].

Test Procedure
An ignition sweep was performed to find the ignition timing for MBT (minimum advance for best torque) minus one percent at each of the 25 test points before the responses were recorded. Four engine test conditions (2500 rev/min 5.5 bar bmep, 1500 rev/min 2.62 bar bmep, 1000 rev/min 1.0 bar bmep, and idle) were investigated to assess engine performance over a range of part-load operation. It was not possible to run at MBT for many of the test points at 2500 rev/min 5.5 bar bmep due to detonation, and so an ignition setting of knock minus two degrees was used in these instances.

Engine Test Results
It may be imagined that that the testing generated a large amount of data even with the limited number of test points, and so only a selection can be included here. The results will be restricted to showing the responses of NOx and uHC emissions as well as burn rate at the 1500 rev/min, 2.62 bar, condition. The results obtained at the 1000 rev/min, 1.0 bar bmep, condition were generally similar in their trends, whereas the influence of knock limitations could be seen to affect the results at the 2500 rev/min, 5.5 bar bmep, condition.

Figure 1 shows the factors that affected the NOx emissions. A high standardised 't' value for a variable indicates a large influence by that variable, with all values greater than 2.1 indicating that the influence of that variable may be considered to be significant. It is apparent that the most important influences are intake camshaft (variable B) and exhaust camshaft (variable A) phasings. The relatively high value of AA indicates that there was a significant non-linearity (ie curvature in the response curve) in the effect of exhaust camshaft phasing on NOx emissions. The column AB indicates that there was significant interaction between the phasings of the intake and exhaust camshafts; in other words, the effect of exhaust camshaft phasing on NOx emissions was different for different intake camshaft phasings, and vice versa.

Even modern computer graphics have not been developed to represent four-dimensional surfaces, so any figures showing the details of the responses must be limited to two or three-dimensional plots. Figure 2 shows how the NOx emissions for one cylinder bank vary with intake and exhaust camshaft phasings with the other two variables, intake camshaft duration and lift, at their mid-point values. Figure 3 is a similar plot showing the effects of intake camshaft duration and intake camshaft phasing, with the other two variables at their mid-point values. It is apparent that combinations of the variables that give a large overlap between intake and exhaust camshafts tended to give low NOx values. This is an expected result since large overlaps will tend to lead to high proportions of residual gas within the cylinder, with a consequent reduction in peak combustion temperature and NOx emissions. More surprising is the indication from figure 1 that intake camshaft lift was not a significant variable, since the expectation is that it would affect burn rate with some consequence on NOx emissions. In fact, the burn rate was calculated for each test point in terms of the 0-80% burn angle for an average of around 150 cycle pressure traces. A simple first-law calculation, as described by Stone [6], was used since it was intended that the results would only be used for comparison one with another rather than as absolute measures. The results shown in figure 4 suggest that there is no one variable or interaction between variables that has a really significant influence on burn rates. The picture is not clear, but it would seem that valve timing events (ie phasings and duration) do have some limited effect. The plots of the variables show that, in general, high overlap conditions give long burn durations, which is consistent with the behaviour that would be expected with a larger proportion of residual gases. Valve lift appears as a DD term with some significance, and as a CD term with a weaker significance. It is revealed in figure 5 that the DD effect results from a characteristic which shows the 5.5 mm peak valve lift to be a watershed in that peak valve lifts both above and below give shorter burn durations. Possible reasons for this are discussed later.

The only variable with a strong significance in its effect on uHC emissions was exhaust camshaft phasing. The plot of the effects of the variables on uHC emissions (figure 6) indicates that retarding the exhaust camshaft phasing reduced uHC emissions. A possible reason is that retarding the exhaust camshafts gives a later closing of the exhaust valves, with consequently a greater chance that the last exhaust gases leaving the cylinder, uHC-rich as they included top-land crevice gases [2], were re-ingested rather than being exhausted to add to the overall uHC concentration in the exhaust. Another explanation is that exhaust temperature was higher because retarding the exhaust camshafts led to a higher proportion of residual gas which, in turn, gave a lower burn rate; the higher gas temperature leading to increased oxidation of uHCs.

Modelling the gas exchange processes
The gas exchange processes were modelled so as to obtain a greater understanding of the engine test results. A 'filling and emptying' model was adopted as described by Watson and Janota [7], with the engine being considered as separate manifold and cylinder volumes connected through the intake valves. A series of rig tests were conducted to measure intake and exhaust valve discharge coefficients for realistic flow rates under both forward and reverse flow conditions. These were combined with the known lift characteristics of the various camshaft designs in equations to predict the flow rate versus time characteristics for flows passed both the intake and exhaust valves. The assumption of perfect mixing between air and any burnt gas flowing back into the inlet manifold was made in order to calculate the proportion of residual gas trapped in the cylinder at the start of compression. This assumption simplified the computation, but means that the computed values of the trapped residual gas fraction (RGF) are likely to be in error in absolute terms, although they should be valid for indicating trends.

It was realised during this exercise that the valve clearance had an important effect on the results for some camshaft phasing and duration combinations. The engine used for testing did not have hydraulic valve lifters, and so specified clearances (0.20 mm for the inlet and 0.25 mm for the exhaust valves) were set at the beginning of each test when the engine was cold. The percentage of RGF was computed for each test point at the 1500 rev/min, 2.62 bar bmep, condition assuming that the clearances of the valves remained at the cold settings. The results for 6 of the 25 CCRD test points are compared for three different valve clearance assumptions in figure 7. Each of the camshafts was designed with a 0.45 mm ramp between the base circle and the start of the lift profile. One set of results, termed "standard valve clearance", was computed assuming that the clearance remained at its cold value, another set assumed that differential expansion increased the cold clearance by 0.1 mm (ie 0.30 mm clearance left before the start of intake cam profile), while a third assumed that the full ramp height provided the clearance (ie 0.45 mm valve clearance). It can be seen that the RGF was changed by up to 8% over the range of valve clearances shown in figure 7. There were also significant changes in the computed mean effective pressure of the pumping loop from one clearance to another (figure 8). These orders of differences can be explained by reference to the computed cylinder pressure versus volume diagram for the gas exchange part of the cycle, as shown in figure 9. The loops shown in figure 9 correspond to test number 1 in figures 7 and 8 for the two extremes of valve clearance. The exhaust valves closed early for this test (15° bTDC) while the intake valves opened at 12.5° bTDC, all valve event timings specified on the assumption of having the full 0.45 mm clearance. The early closing of the exhaust valves has led to extra residual gas being left in the cylinder, particularly so with the largest clearance, as this tends to accentuate this effect. This timing may be one that would not be adopted on an actual engine, but it is within the range selected for the CCRD experiment in order to build up full response surfaces. After due consideration, it was decided to work with the values calculated from the model assuming the standard valve clearances.

The computed values of RGF for the standard valve clearances were compared to both NOx level (figure 10) and 0 - 80% burn duration (figure 11). It can be seen that in both cases the expected trends are apparent, indicating that the probable mechanism by which the variables affected NOx and burn duration was through their influence on RGF. The scatter in the points is due to a combination of factors, namely experimental inaccuracies, assumptions made in the computation of RGF, and secondary influences such as changes in peak lift influencing in-cylinder motion and hence combustion.

Rig test results

Steady-flow tests were performed on a tumble-measuring test rig based on a cylinder head of the same design as that fitted to the test engine. The objective was to understand how changes in the peak valve lift might affect in-cylinder motion. A standard configuration of rig [8] was used to measure tumble, the arrangement shown in figure 12. The tumble meter consisted of an extruded aluminium section containing a series of narrow passages. The reaction exerted by the flow section in straightening the swirling flow gave a measure of the tumbling motion created in the "dummy" cylinder. Features of the rig were the setting of the false piston at the depth corresponding to the bottom dead centre position of the piston on the engine, and the inclusion of twin balanced pipes for the air draw-off to the vacuum pump. The results are given in figure 13 in terms of tumble coefficient which is defined as:

$$\text{Tumble Coefficient} = \frac{8 \times T_r}{m \, v_c \, B}$$

where T_r = torque reaction on flow straightening section
 m = mass flow rate of air
 v_c = velocity of air at valve curtain area
 B = cylinder bore

The testing was performed at a wide open throttle setting as this is the normal test condition used by industry for tumble measurements. Virtually no tumbling motion was measured for the first few millimetres of valve lift. This could provide an explanation for the influence of valve lift on burn duration as shown in figure 5, where burn duration is shown to reduce for peak lifts either side of 5.5 mm. A hypothesis to explain this behaviour is that burn duration is shortened at the higher peak lifts by the establishment of a useful tumbling motion within the cylinder. On the other hand, at low peak lifts, there is no significant tumble generated but the higher velocities of the intake charge do generate some beneficial turbulence, albeit rather early during the compression process.

Summary and Conclusions

1   A comprehensive series of tests have been performed to investigate the effects of intake and exhaust camshaft phasings, intake camshaft duration, and intake valve lift on engine-out emissions and other performance parameters of a prototype version of a production 4-litre, Vee-8, engine with 4 valves per cylinder. A statistical approach was adopted to the design of the experiments and the analysis of the results. A Central Composite Rotatable Design of experiment was applied as this reduced the number of test points to 25 for each of the 4 part-load engine conditions investigated. This approach was successful in identifying the main effects and second-order interactions.

2   It was apparent that the main influence on oxide of nitrogen (NOx) emissions was any alteration of the valve timing events that led to changes in the residual gas fraction (RGF) trapped in the cylinder at the start of compression. The RGF had a consequent effect on peak combustion temperature and hence NOx formation.

3   It was somewhat surprising that the peak lift of the intake camshaft did not have a significant influence on NOx emissions. It might have been expected that peak lift would affect burn duration and hence NOx formation.

4   Investigation of burn duration showed some influence of valve lift, such that a peak lift of 5.5 mm gave the longest duration of burn, with shorter durations at both lower and higher lifts.

5   A model of the gas exchange processes revealed that valve clearances have a surprisingly significant effect on RGF and pumping loop mean effective pressure for some camshaft phasings and durations. It is an influence that would tend not to be encountered on engines where no mechanism is fitted to alter any of the valve events but, as more valve phasing and profile switching systems are adopted, it should be recognised as a possible source of variability in engine behaviour.

6   Steady-flow rig testing showed that virtually no tumble was induced within the cylinder at low values of intake valve lift, but tumble did increase with valve lift. This provided a possible explanation for the shorter burn durations of valve lifts either side of 5.5 mm peak lift. Tumble would lead to shorter burn durations for the higher peak lifts, whereas the increased velocities of intake flows at low peak lifts would induce extra turbulence and reduce burn duration.

Acknowledgements

The work was supported by the Engineering and Physical Sciences Research Council through a CASE award, with further support from Jaguar Cars, and this is most gratefully acknowledged. The authors would like to thank the following for their valuable help: Mr Chris Atkinson, Mr Xiang Dong Chen, Mr Martin Joyce, Mr Gordon Leeming, and Mr Stuart Woodcock all of Jaguar Cars for advice and support during the engine test programme, and also Mr Terry Curtis of UCL for technical assistance throughout the project. Thanks are also due to Mr Paul Bennett of BP for arranging and advising on the supplies of fuel throughout the testing period, and Mr Greg Allin of Signal Instruments for loan of gas analysis equipment at times of need. The earlier work of Dr Justin Seabrook (Cosworth Technology) during his time at UCL led to the development of some of the programs used for the analysis of the engine results, and this is duly acknowledged.

References

1   Leone, T G, Christenson, E J, and Stein, R A,   "Comparison of Variable Camshaft Timing Strategies at Part Load", SAE 960584.

2   Seabrook, J, Nightingale, C, and Richardson, S H,   "The Effect on Hydrocarbon Emissions - An Investigation with Statistical Experiment Design and Fast Response FID Measurements", SAE 961951.

3   Horie, K, Nishizawa, K, Ogawa, T, Akazaki, S, and Muira, K,   "The Development of a High Performance Four-Valve Lean Burn Engine", SAE 920455.

4   Infinitely variable valve control - almost the best of all worlds, Paramins Post Industry and Additive News, Issue 3, Exxon Chemicals Ltd., London, UK.

5   Wilson, N D, Watkins, A J, Dopson, C,   "Asymmetric Valve Strategies and their Effect on Combustion", SAE 930821.

6   Stone, R, "Introduction to Internal Combustion Engines", Second Edition, Macmillan Press, 1992.

7   Watson, N, and Janota, M S,   "Turbocharging the internal combustion engine", Ellis Horwood, 1984.

8   Chapman, J, Garrett, M W, and Warburton, A,   "A new standard for barrel swirl measurement", IMechE Autotech Conference, Section 18, Mechanical Engineering Publications Ltd, 1991.

Test Case No.	EVO °bBDC	IVO °bTDC	EVC °aTDC	IVC °aTDC	Peak Lift (mm)
1	65	12.5	-15	202.5	4.4
2	35	12.5	15	202.5	4.4
3	65	-17.5	-15	232.5	4.4
4	35	-17.5	15	232.5	4.4
5	65	27.5	-15	217.5	4.4
6	35	27.5	15	217.5	4.4
7	65	-2.5	-15	247.5	4.4
8	35	-2.5	15	247.5	4.4
9	65	12.5	-15	202.5	7.2
10	35	12.5	15	202.5	7.2
11	65	-17.5	-15	232.5	7.2
12	35	-17.5	15	232.5	7.2
13	65	27.5	-15	217.5	7.2
14	35	27.5	15	217.5	7.2
15	65	-2.5	-15	247.5	7.2
16	35	-2.5	15	247.5	7.2
17	80	5	-30	225	5.8
18	20	5	30	225	5.8
19	50	35	0	195	5.8
20	50	-25	0	255	5.8
21	50	20	0	240	5.8
22	50	-10	0	210	5.8
23	50	5	0	225	3.0
24	50	5	0	225	8.6
25	50	5	0	225	5.8

* Test 25 is the Centre Point (repeated 7 times)

Table 2   Valve events for 25 test points

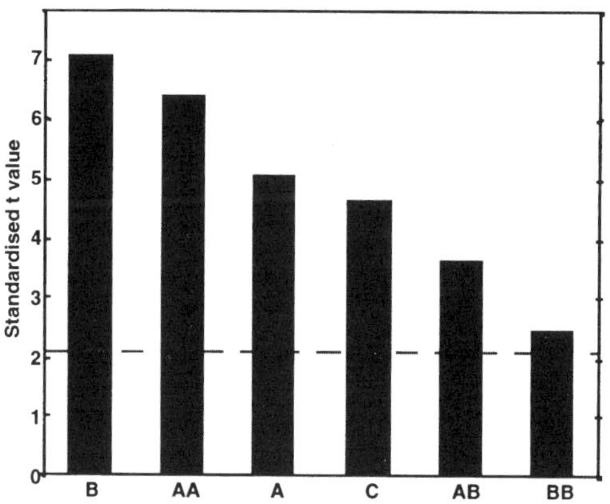

Dashed line represents lowest standardised t value for which the effect of a variable may be considered significant

A   Exhaust phasing
B   Intake phasing
C   Intake Duration
D   Intake lift

Figure 1   Factors influencing NOx emissions at 1500 rev/min, 2.62 bar bmep

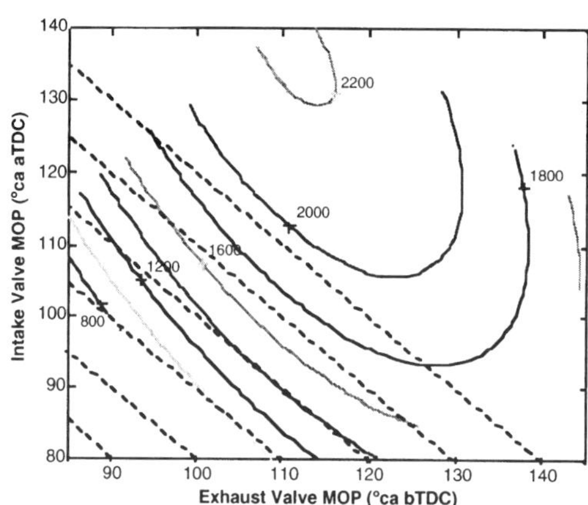

Values of NOx emissions in ppm

Dashed lines correspond to lines of constant valve overlap (highest overlap in bottom left hand corner)

(MOP is mid opening point)

Figure 2   Influences of intake and exhaust camshaft phasings on NOx emissions at 1500 rev/min, 2.62 bar bmep

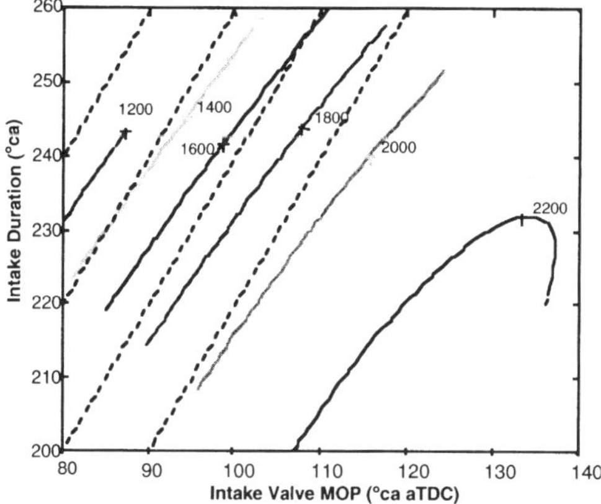

Values of NOx emissions in ppm

Dashed lines correspond to lines of constant valve overlap (highest overlap in top left hand corner)

(MOP is mid opening point)

Figure 3   Influences of intake camshaft phasing and intake camshaft duration on NOx emissions at 1500 rev/min, 2.62 bar bmep

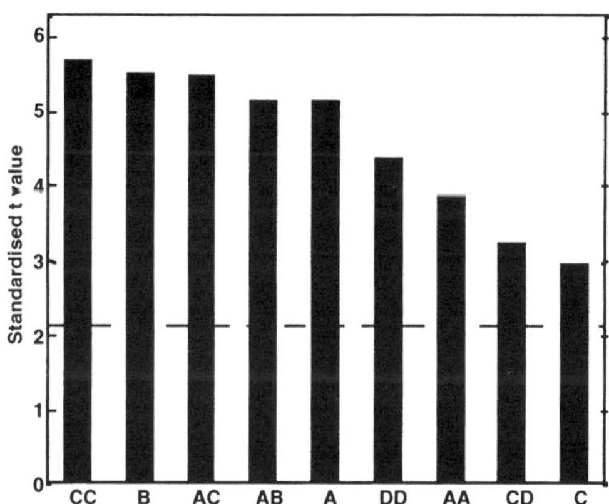

Dashed line represents lowest standardised t value for which the effect of a variable may be considered significant

A	Exhaust phasing
B	Intake phasing
C	Intake Duration
D	Intake lift

Figure 4   Factors affecting 0 - 80 % burn duration at 1500 rev/min, 2.62 bar bmep

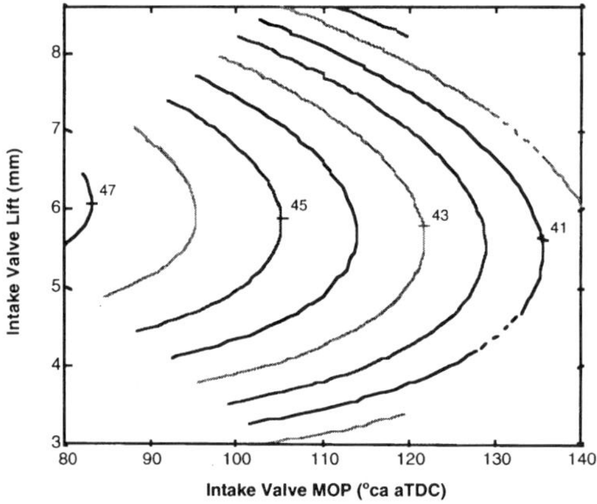

Figure 5   Influences of intake valve lift and intake valve phasing on 0 - 80% burn duration at 1500 rev/min, 2.62 bar bmep

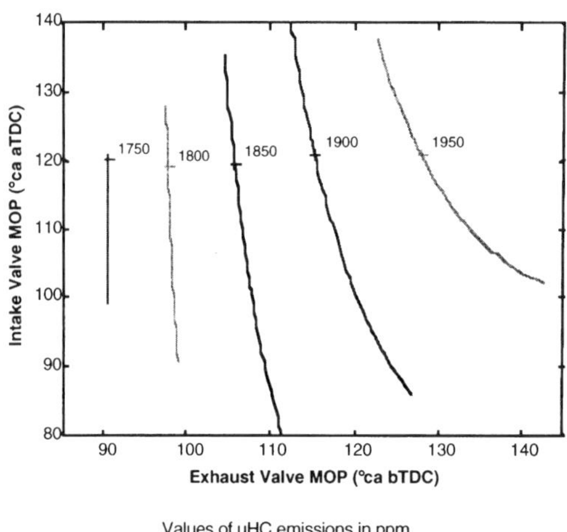

Figure 6   Influences of intake and exhaust camshaft phasings on uHC emissions at 1500 rev/min, 2.62 bar bmep

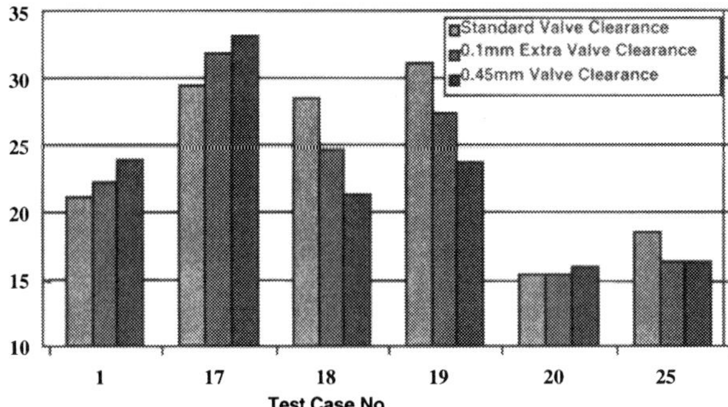

Figure 7   Predicted percentage residual gas fraction (vertical axis) trapped in the cylinder for six test points, 1500 rev/min, 2.62 bar bmep

Test Case No.	EVO (°ca bBDC)	IVO (°ca bTDC)	EVC (°ca aTDC)	IVC (°ca aTDC)	Peak Lift (mm)
1	65	12.5	-15	202.5	4.4
17	80	5	-30	225	5.8
18	20	5	30	225	5.8
19	50	35	0	195	5.8
20	50	-25	0	255	5.8
25	50	5	0	225	5.8

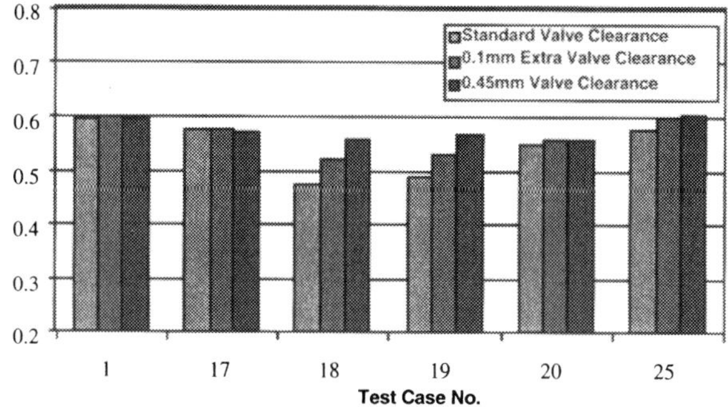

Figure 8   Predicted pumping mean effective pressure in bar (vertical axis) for six test points, 1500 rev/min, 2.62 bar bmep

Test Case No.	EVO (°ca bBDC)	IVO (°ca bTDC)	EVC (°ca aTDC)	IVC (°ca aTDC)	Peak Lift (mm)
1	65	12.5	-15	202.5	4.4
17	80	5	-30	225	5.8
18	20	5	30	225	5.8
19	50	35	0	195	5.8
20	50	-25	0	255	5.8
25	50	5	0	225	5.8

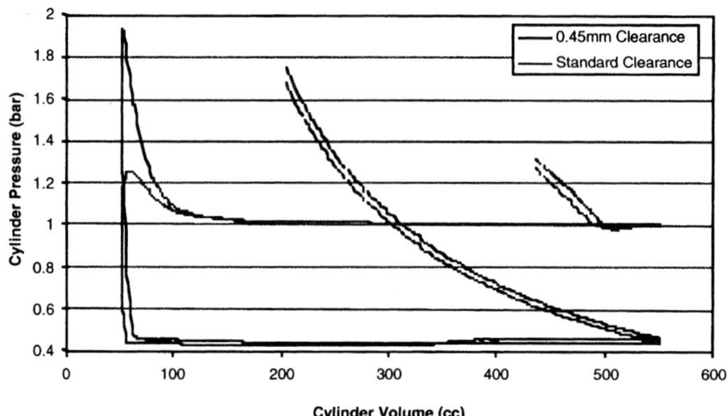

Figure 9 Predicted pumping loop for standard and 0.45 mm valve clearances at 1500 rev/min, 2.62 bar bmep   (Test Case No. 1)

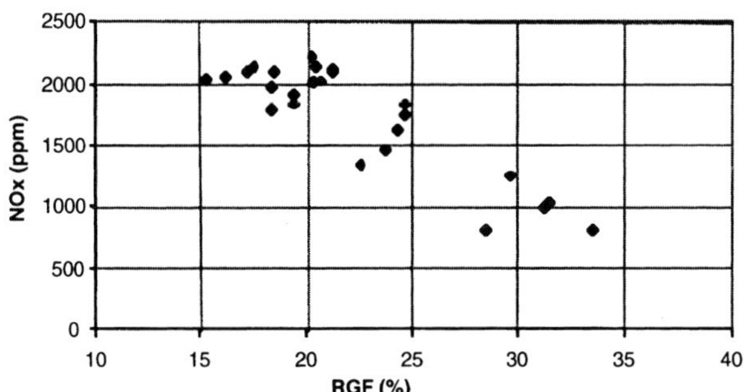

Figure 10 Measured NOx emissions versus predicted percentage residual gas fraction (standard valve clearances) at 1500 rev/min, 2.62 bar bmep

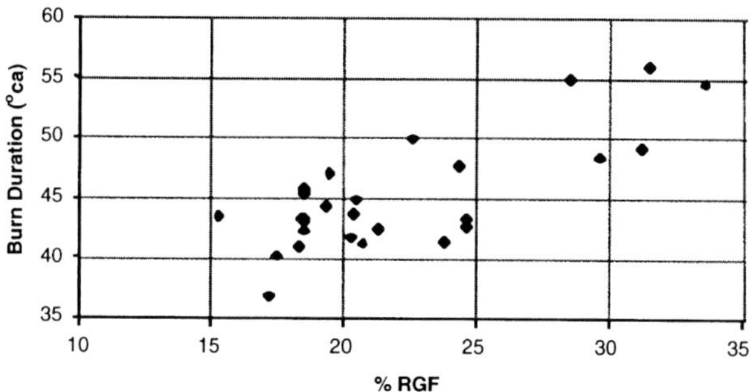

Figure 11   0 - 80% burn duration (crank angle degrees) versus predicted percentage residual gas fraction at 1500 rev/min, 2.62 bar bmep

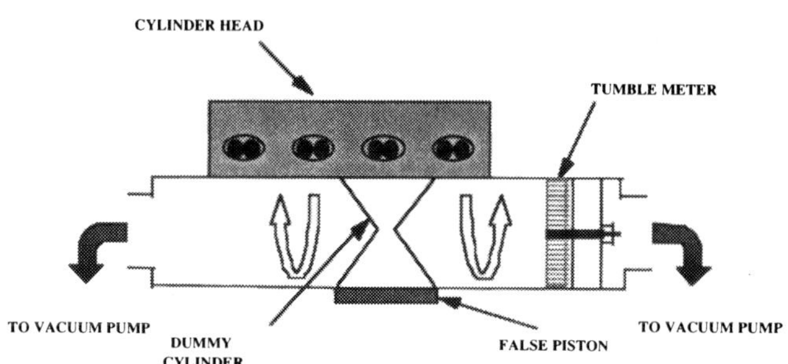

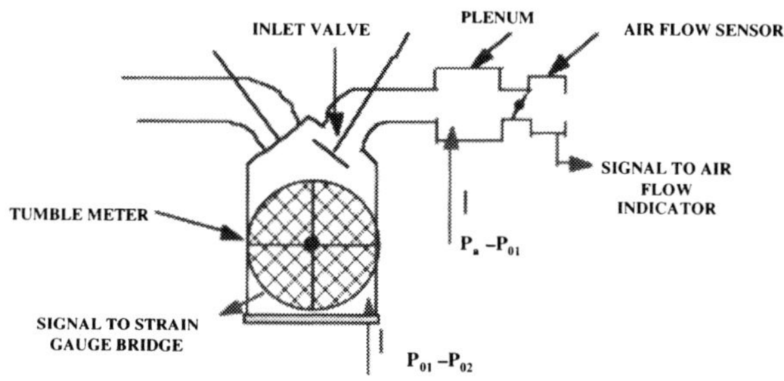

Figure 12   Schematic layout diagrams of tumble measuring rig

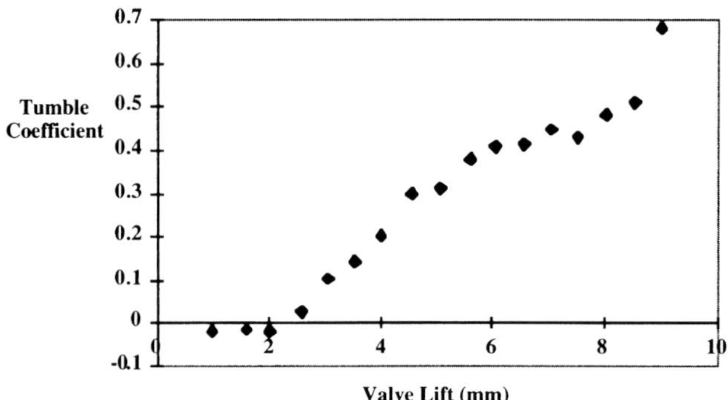

Figure 13   Tumble coefficient versus versus valve lift (mm) - unthrottled

The analysis of an engine with a variable valve train

A FROMMELT, HEEL, LIANG, SCHULZE, and **WUNDERLICH**
Daimler Benz

Abstract:
The stress collective of a high performance petrol engine in its application as a passenger car drive is characterised by high partial load proportions. If the load control of the petrol engine is achieved through throttling under these lower load range operating conditions, there are very high gas exchange losses (throttling losses), resulting thereby in a specifically poor, system-dependent partial load consumption.

Recently increasing efforts have been undertaken with different systems, such as cylinder disconnection, multi-valve technique, channel disconnection, camshaft adjustment or the variable valve train, to lower the emission and the fuel consumption while enhancing specific engine performance at the same time.

Through thermodynamic analyses, the engine improvement potential is demonstrated with the help of the variable valve timing system on the basis of test results from a single-cylinder unit concerning fuel consumption and performance.

This paper provides further details on different methods of technical implementation of variable valve gears, the pros and cons of the systems as compared to an engine with a conventional valve system and the thermodynamic or specific flow-related features of the engines with variable valve impulse. Of special interest here is the analysis of throttle-free load via an advanced variable intake close and maximum stroke reduktion

The test unit is a single-cylinder engine, the variation of the different valve timing is re-alised with the help of a hydraulically actuated computer-controlled unit. The tools applied to analyse the engines were a combustion and load change analysis as well as a three-dimensional flow calculation.

1 Introduction

Different variable valve control methods, with whose control time curves throttle-free load control methods can be realised, were examined in extensive basic investigations conducted on a computational and experimental basis /1/. Various methods were investigated while taking into consideration the technical requirements, the raw exhaust gas emissions and the achieved reduction in the gas exchange requirements. Method 1 implements load control with nearly constant inlet valve open, in connection with variable inlet valve closed and a valve lift which depends on that. The second method enables load control with constant inlet valve closed but constant maximum valve lift. The position of the valve overlap here is adapted to the respective load point through internal exhaust gas recirculation according to the exhaust gas compatibility.

Both methods were examined on the single-cylinder unit and thermodynamically analysed; method 1 was documented in the attachments.

2 Description of the test unit

Figure 1 shows the schematic structure of the test unit with fully variable valve control. Virtually any outlet and inlet valve lift curves can be set with the single-cylinder unit and the adapted hydraulic Lotus valve control. The valve lift curves in the exhaust stage were kept constant according to a series cam shape, while the inlet valve lifts and control times, on the other hand, were varied according to strategy in order to implement the function of load control with reduced gas exchange requirements and with the help of variable opening cross-sections.

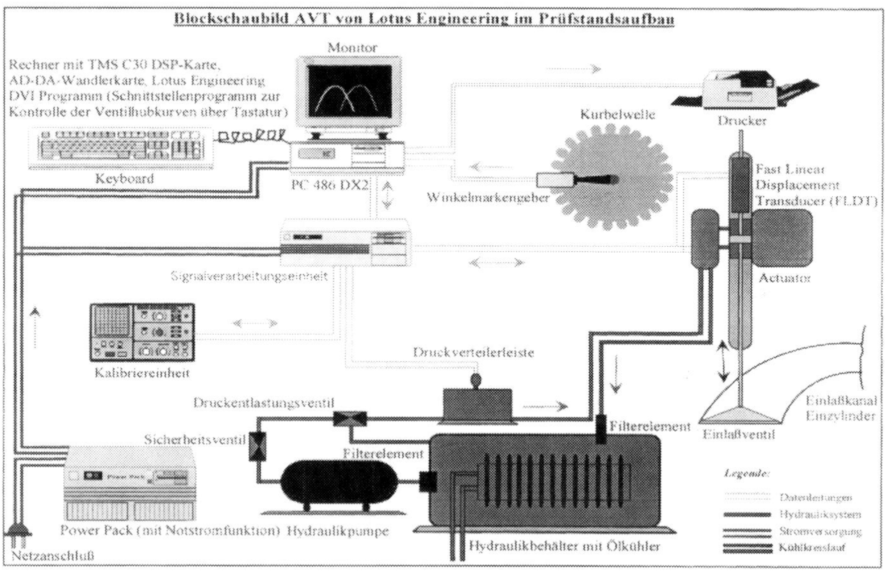

Figure 1: Hydraulic valve control of test unit

Based on this test unit used exclusively for research purposes, the following variable valve control methods, some in the prototype phase, were investigated with regard to possible implementation.

- VLC   Variable lift control system (mechanical valve drive)
- EMVC   Electro-magnetic valve control
- EHVC   Electra-hydraulic valve control

3 Thermodynamic principles

The greatest difference between an aspirating engine with conventional valve control and an engine with variable valve control relates to the low-loss gas exchange under partial load. Due to the comparably homogeneous mix preparation of the concepts examined, the differences under full load conditions are small with respect to combustion, but substantial differences emerge in the attainable cylinder filling.

The operating point at 2000 rpm has been analysed in detail and is documented in the figures at an effective mean pressure of 2 bar. The basic thermodynamic effects and effective mechanisms are similar in the three concepts mentioned, i.e. VLC, EMVC and EHVC; the following analyses are based on the VLC system.

The entire fuel energy fed to the engine (Fig. 2) was converted to the mean pressure unit and divided into effective and loss portions. The calculated mean frictional pressure or degree of mechanical efficiency is of minor importance because of the variations in frictional loss typical of the unit and it appears sensible to compare in the further analysis on the basis of indicated variables.

An assessment of the concepts must take into account that the work for valve actuation of the externally hydraulically driven Lotus unit is not contained in the energy balance.

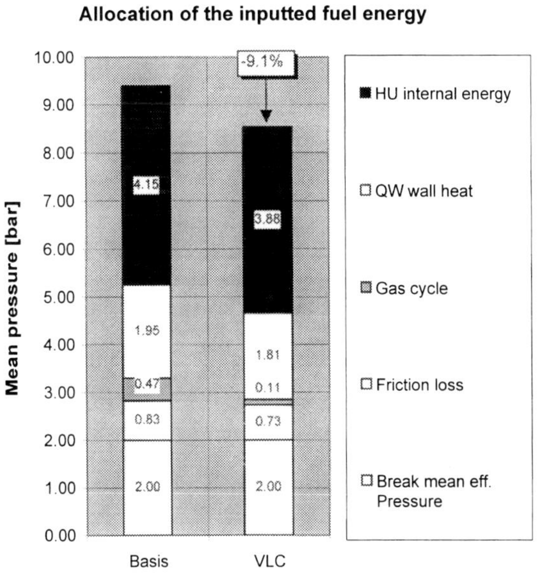

Allocation of the inputted fuel energy

Fig. 2: Energy balance

3.1 Gas exchange

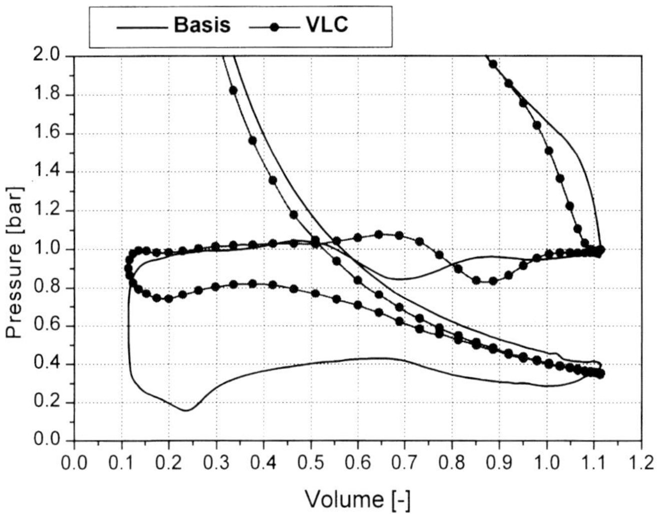

Fig. 3: Comparison of gas exchange loops

The area of the gas exchange loop in the p-v diagram is a measure of the pumping work. The lower pumping work was measured for the unit with variable valve drive.

	Basis	VLC
Pumping work **[bar]**	0.47	0.11
Gas exchange efficiency [%] $\eta_{LP} = \dfrac{IMEP\_HP - \lvert IMEP\_LP \rvert}{IMEP\_HP} [\%]$	85.7	96.0

Based on the measured pressure curves in the cylinder, manifold and exhaust pipe (in immediate proximity to the valves in each case), the gas exchange analysis provided data on the time curve of the mass flows in the inlet and outlet valve, the concentration of residual gas, return flow effects into the manifold and the potential of control manifolds in connection with variable control times.

The time curve of the calculated mass flows is compared in Figures 4 and 5. Due to the substantially reduced effective valve opening cross-section and the fact that the opening time of the inlet valve has been cut nearly in half, the mass flow curve for the VLC unit at this operating point is more even and has no return flow effects, apart from the overlap phase. The calculation of the residual gas portions resulted in 14% for the throttle-regulated basic unit and 17% for the VLC unit.

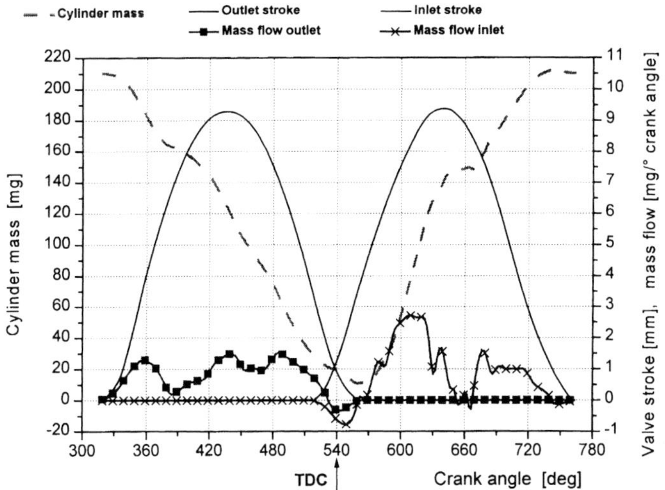

Fig. 4: Mass flows of basic unit

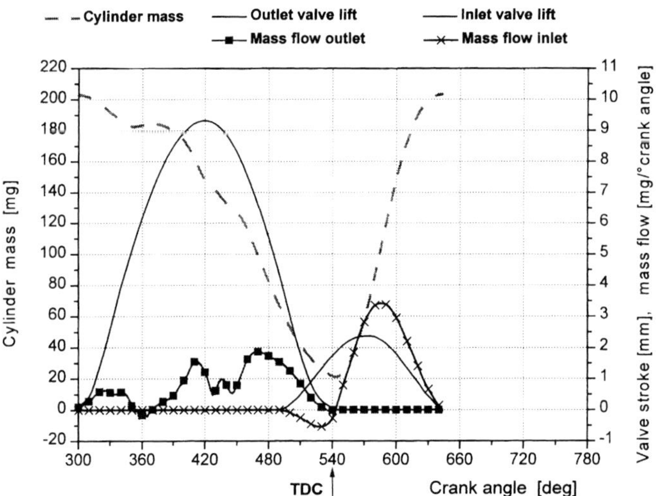

Fig. 5: Mass flows of VLC unit

3.2 Combustion

The calculated combustion curves (Fig. 6) show significant differences in the form and duration of combustion. The location of 50% mass fraction burned is set to an optimal degree of efficiency at an 8-degree crank angle after ignition TDC with both combustion methods.

The duration of the combustion is considerably longer for the unit with variable valve drive, based on the basic unit run without external exhaust gas recirculation.

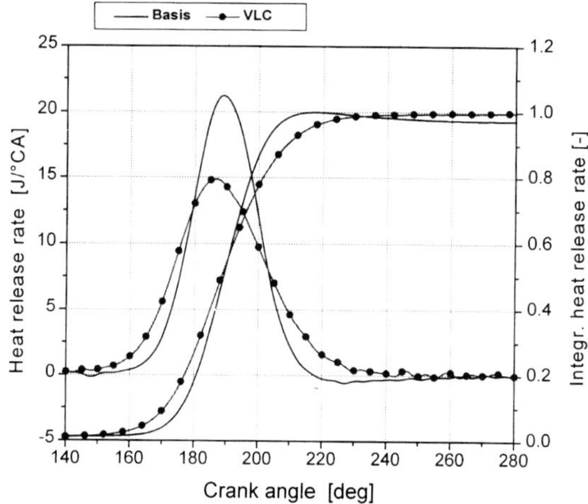

Fig. 6: Heat release rate

One cause of the increasing duration of combustion is the rise in the residual gas concentration through the "internal exhaust gas recirculation" (see gas exchange analysis in section 3.1).

Section 3.3 will deal with other causes, such as mix homogenisation and gas temperature, in more detail.

The IMEP HP (high pressure) cycle scatter shown in Fig. 7, applied above the location of 50% mfb, shows that the scatter of the 50% mfb of successive working cycles is more pronounced in the VLC concept.

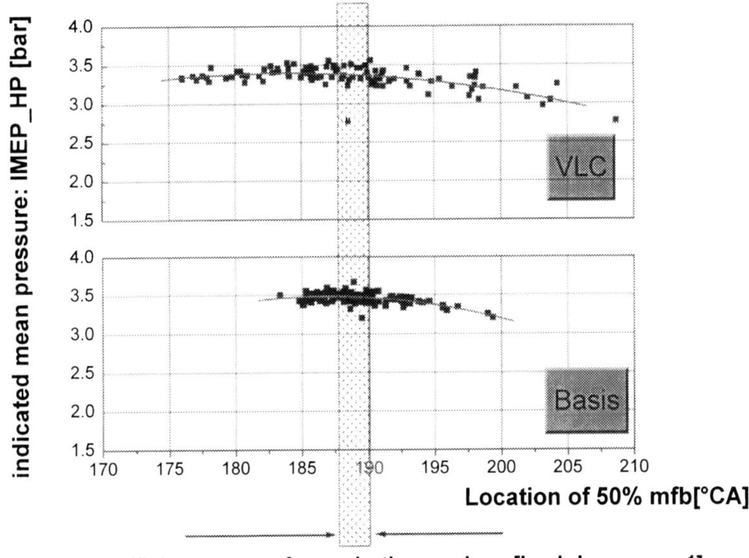

optimal efficiency range for aspirating engines [lambda approx. 1]

Fig. 7: Scatter of combustion

For an overall assessment of the combustion and gas exchange, without any corrupting influence of various frictional factors, the indicated efficiency is the correct assessment variable. The indicated efficiency is the product of the high pressure efficiency and the gas exchange efficiency. Based on the basic unit, an improvement in the indicated of efficiency of 7 percent resulted for the unit with variable valve drive.

High-pressure efficiency [%]	Basis	VLC
$\eta_{HP} = \dfrac{IMEP\_HP}{IMEP\_FUEL}[\%]$	35.2	33.4
Indicated efficiency [%]		
$\eta_{IMEP} = \eta_{HP} * \eta_{LP}[\%]$	30.0	32.1

3.3 Load control

On the basis of load control via the throttle valve, this test series examined various intermediate stages ranging to sole load control via the inlet valves. This test series represents a gradual transition from the conventionally throttled aspirating engine to the "VLC aspirating engine" (inlet valve lift = 1.8 mm, manifold partial vacuum = 0 mbar), see Fig.8.

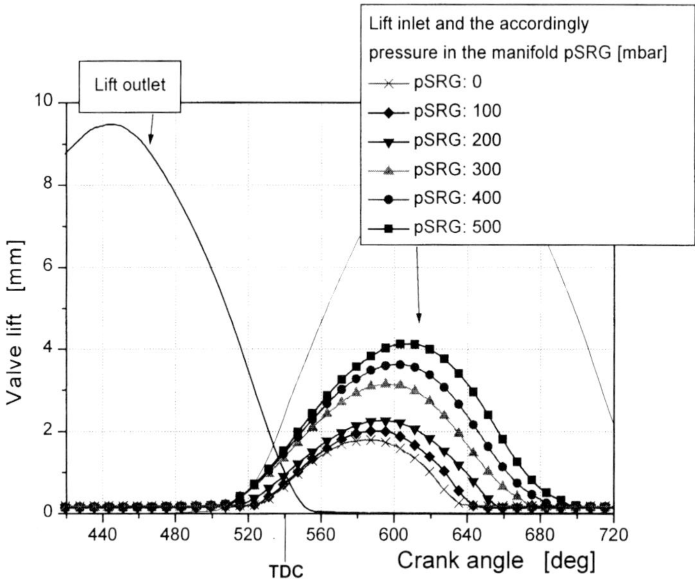

Fig. 8: Load control via inlet valve lift nMot: 2000 rpm pme: 2 bar

The advantages of load control via the inlet valves are significant with respect to the gas exchange required. The IMEP_LP (low pressure) requirement is reduced from 0.42 to 0.11 bar (Fig. 9) in this test series.

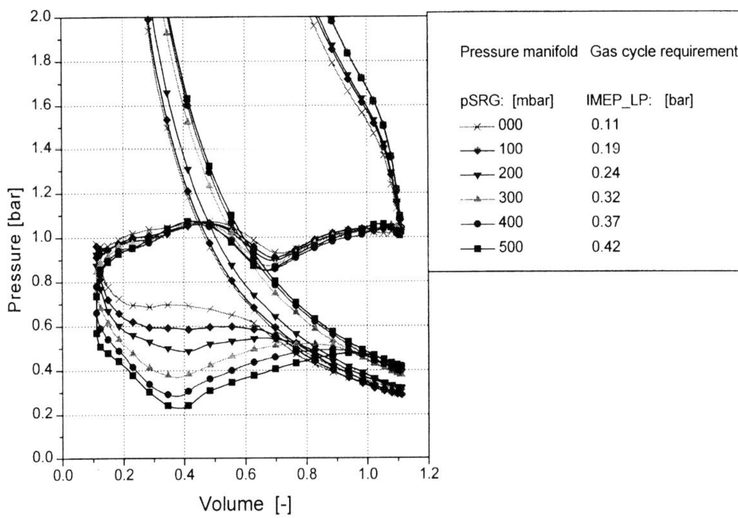

Fig. 9: Effects on gas exchange requirement

Due to the early closing of the inlet valve of up to an 80-degree crank angle before bottom dead centre and the following expansion of the fresh gas until bottom dead centre, the greatest vacuum and thus the lowest temperature in the case of the VLC concept results at the time of BDC. In the event of a very early closing of the inlet valve and thus pronounced expansion, the fresh gas taken in cools down by up to 40 K (Fig. 10). During the subsequent compression this cooling leads to unfavourable conditions for the mix preparation and combustion.

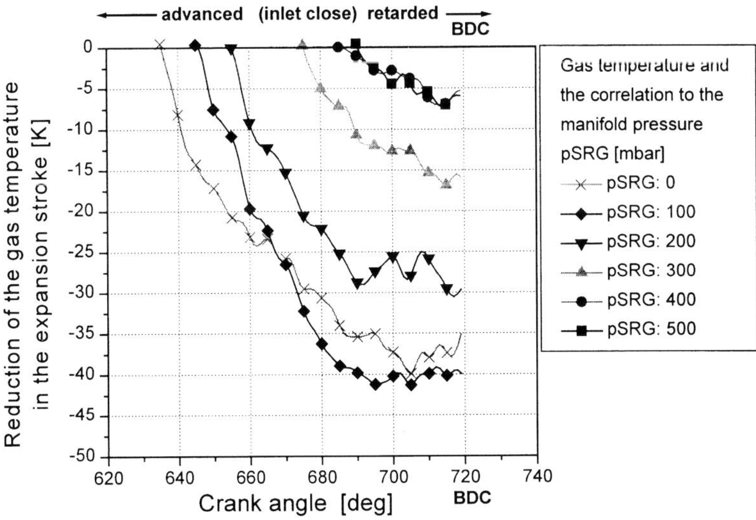

Fig. 10: Gas temperature in the expansion stroke

This is shown by a preignition requirement that is up to 200 (crank angle) larger and by a significantly longer mean duration of combustion (Fig. 11).

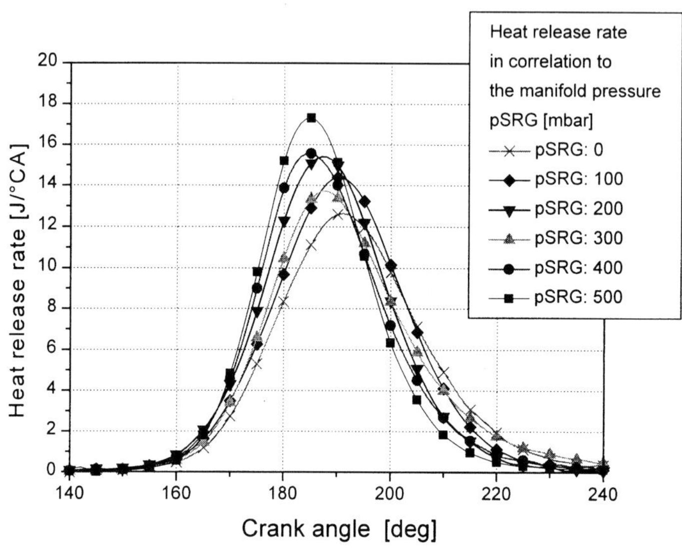

Fig. 11: Effect on combustion curves

In a cross comparison (Fig. 7) the scatter of the location of 50% mfb is largest in the VLC unit.

The following have a negative influence on the duration and stability of the combustion

- Temperature drop due to expansion
- Restricted charge movement due to early closing of inlet valve
- Early ignition point – lower compression temperature – increasing ignition delay

In a project of the Research Association for Internal Combustion Engines (FVV) /2/ the causes for the poorer combustion with throttle-free load control through early closing of the inlet valve was examined with the help of thermodynamic analyses and schlieren photographs. The intake-side mix formation with throttle-free operation is characterised by fuel deposits on the port walls and valve disc. During the shortened intake phase this fuel does not vaporise completely, is poorly prepared when it reaches the combustion chamber in some cases and wets parts of the combustion chamber surface.

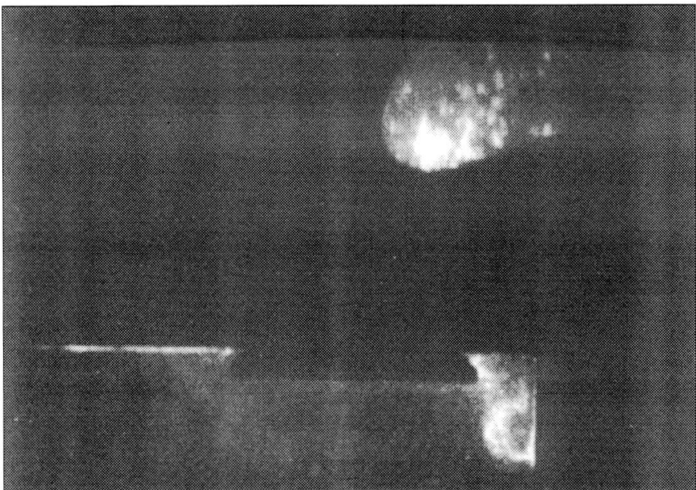

Fig. 12: Schlieren photograph after opening of inlet valve with upstream injection /2/

With help of schlieren photographs, Figure 12 shows the lack of homogeneity of the mix when the inlet valve closes early. The valve seat, valve disc and the nearby cylinder wall are wet with liquid fuel. In some cases the liquid fuel does not vaporise due to the partial vacuum until it reaches the combustion chamber and additionally withdraws heat through the vaporisation energy of the cylinder charge and consequently recondensation of the already vaporised fuel may occur around bottom dead centre.

The mix formation when the inlet valve closes early can be positively influenced through several measures.

- Measures on the injection unit (air-encompassed injection jet)
- Easy throttling in manifold
- Tumble or swirl (charge movement)
- Increase in residual gas concentration (internal, hot exhaust gas recirculation)

The result of the combustion analysis shows that as unthrottling increases, the calculated maximum combustion temperature and the measured peak pressure decline while the exhaust gas temperature rises simultaneously (Fig. 13).

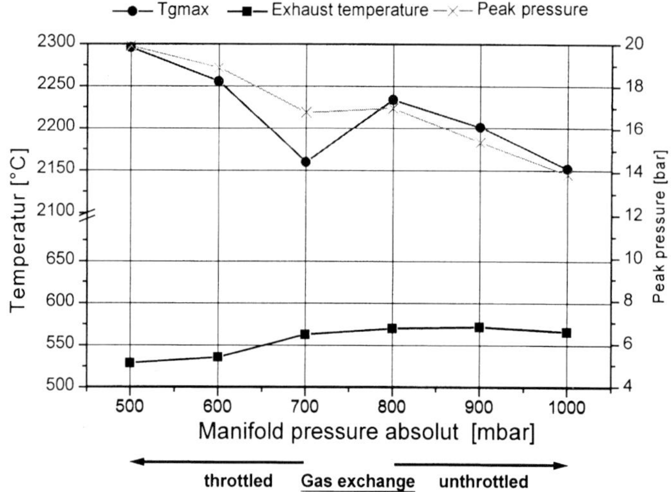

Fig. 13: Temperatures and pressures as a function of throttling

The energy flows calculated from the internal energy balance correlate in accordance with these interrelationships. The increasing proportion of exhaust gas energy or the declining wall heat flow is a consequence of the increasing duration of combustion and is overall the cause of the poorer high pressure efficiency. However, the improvements in the gas exchange efficiency predominate substantially (Fig. 14).

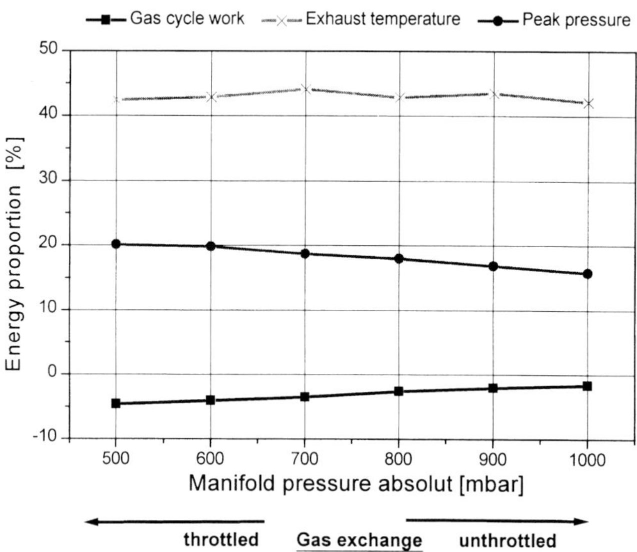

Fig. 14: Energy proportions as a function of throttling

As implied, given insignificantly higher pumping losses, the excessive cooling of the cylinder charge taken in can be avoided through moderate throttling via the so-called exhaust gas actuator.

The manifold pressure fluctuations that result are depicted in Fig. 15. Through this throttling in the manifold the exhaust gas portion flowing back into the manifold during the valve overlap phase is increased. As a consequence, a better mix preparation, lower preignition requirements and less cycle fluctuation result, as expected.

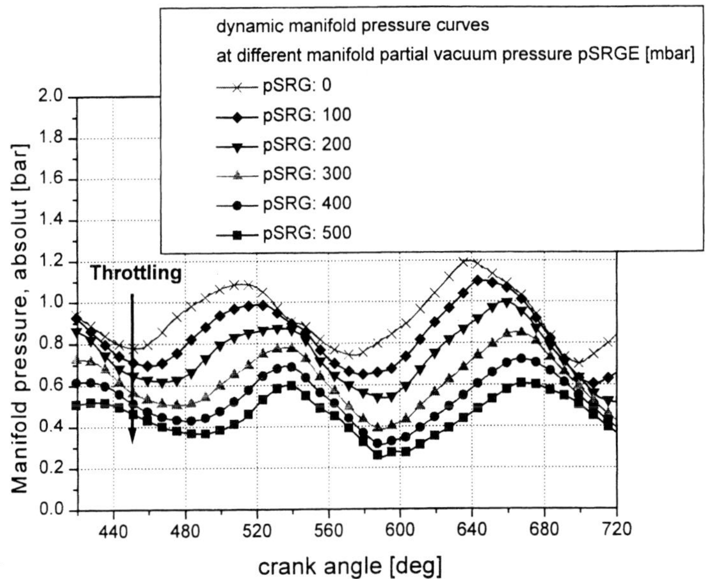

Fig. 15: Throttling

4 Simulation of cylinder interior flow

Examples of results of the three-dimensional flow simulation are shown here to supplement the zero-dimensional analysis of the gas exchange presented in section 3.1. The pressure indicating in the intake and exhause manifold and static temperature measurements are used as a boundary condition. The measured pressure curve was compared to the calculated cylinder pressure curve as a check.

The contribution of the 3D simulation ranges from characterisation of the speed, turbulence, fresh-air and residual gas distribution in the combustion chamber to time–resolution curves of global flow variables, such as tumble count, mean turbulence intensity and mean fresh-gas and residual gas concentration in the cylinder. The local distribution of flow speeds is shown in different views at 90° crank angle after gas exchange TDC in Figures 16-17. The intensive tumble movement that takes place is clearly visible (Fig. 16). The swirling flows directed against each other in the symmetric halves of the cylinder overlie this rotational movement around a horizontal axis of rotation (Fig. 17).

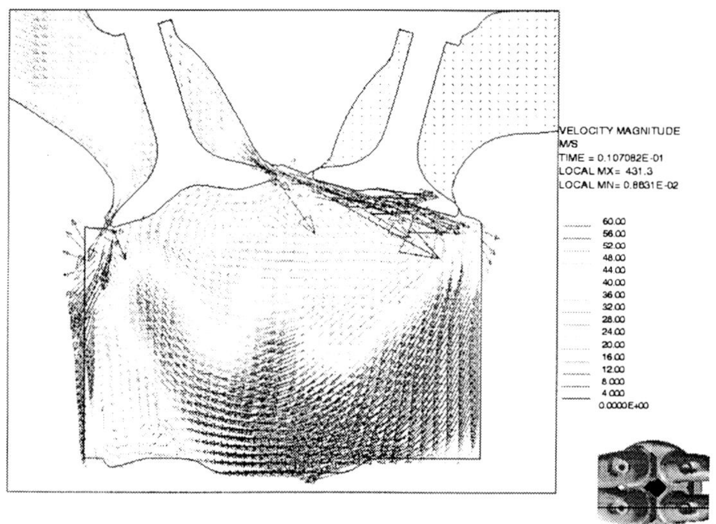

Fig. 16: Speed distribution in vertical section through the axes of inlet valve (left) and exhaust valve (right).

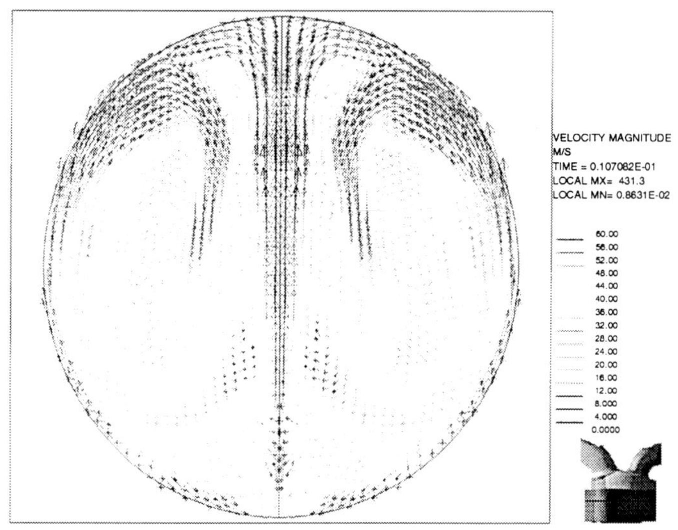

Fig. 17: Speed distribution in horizontal middle section between the piston bottom and the piston top dead centre position.

In addition to the thermodynamic conditions in the cylinder, the level and distribution of turbulence intensity as well as the fuel and/or fresh-air / residual gas concentration at the time of ignition are of crucial importance for combustion. The local distribution of the fresh-air concentration is shown in Figures 18 and 19. For the sake of simplification, a ideally premixed fuel/air mix was specified in the calculation as the initial and boundary condition in the inlet port.

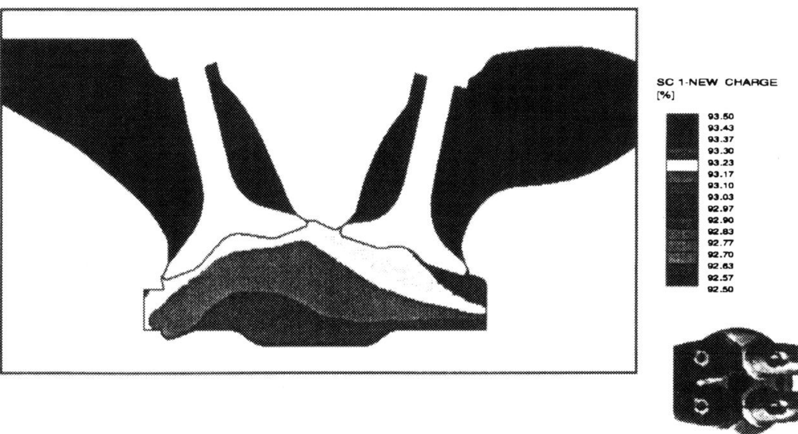

Fig. 18:   Fresh-air concentration distribution in a vertical section through the inlet/exhaust valve plane

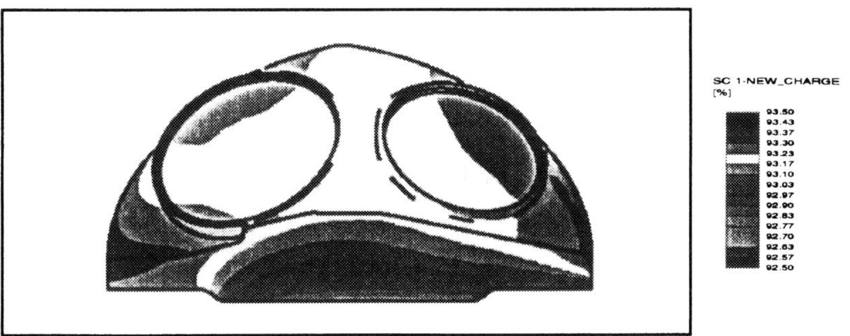

Fig. 19:   Fresh-air concentration distribution in symmetrical half of combustion chamber

Fig. 20 shows the time curves of the turbulent kinetic energy, for which the mean mass was calculated plotted against the combustion chamber, the tumble count as well as the swirl number from inlet valve open to ignition TDC, calculated for a symmetrical half of the combustion chamber. The substantial tumble that forms up to the middle of the intake phase deteriorates into small turbulent eddy structures and thus forms the basis for a high level of turbulence that persists up to ignition TDC. The level of the residual gas concentration has a decisive effect on the laminar flame speed; the turbulent combustion speed is dictated above all by the turbulence intensity.

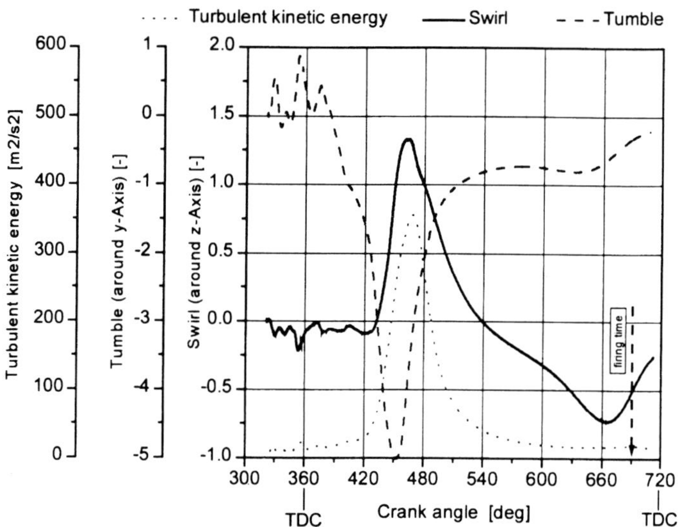

Fig. 20:   Tumble, swirl and mean turbulent kinetic energy in the intake and compression phase

5   Examples of structural implementation of variable valve control systems

5.1   Mechanical variable valve control

This mechanical valve drive system enables control of valve lift and duration of opening of the inlet valves, where the valve lift is always related to a certain duration of opening. This concept involves variable valve control with 2 camshafts for the inlet valves (see Fig. 21) and a conventional camshaft for actuation of the exhaust valves. Control of the inlet valves is provided for via two camshafts that rotate in opposite directions and are connected by means of camshaft phasing (coupler mechanism). Due to the twisting of the two camshafts different lift curve result from the overall kinetics and these curves are transmitted to the rocket via a set of rollers through positive locking. The lift curve family, see Fig. 8, is a function of the shape of the two camshafts, the rocker geometry, the design of the set of rollers and the position of the coupler mechanism.

Fig. 21:   Mechanical variable valve control

5.2   Electromagnetic variable valve control

In the case of electromagnetic valve control, the valve is kept in the open or closed position by means of opening and closing magnets from a half-open neutral valve position that is specified by means of a set of springs. During the valve lift the lift loss caused by frictional losses is compensated for by the electromagnets. The dimensioning of the set of springs must take into account that adequate dynamics are reached at the rated speed and reliable opening of the exhaust valve can be displayed against the combustion chamber pressure. At the time of the opening of the exhaust valve, 40–60° crank angle before lower dead centre, combustion chamber pressures of 7-10 bar must be taken into consideration under full-load operation. When the engine is started, the valves must be guided to a defined closed or open position before or during the starting operation.

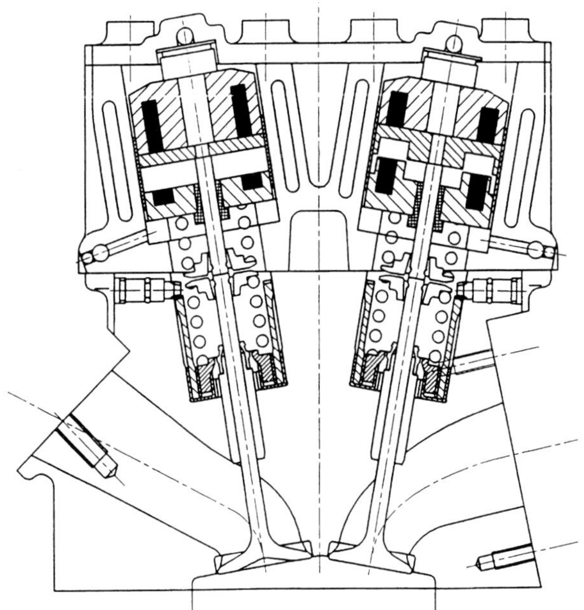

Fig. 22:   Electromagnetic variable valve control

This defined open or closed valve state can be achieved through a single or multiple transient condition. The basic current curve is shown in a schematic diagram in Fig. 23 for the typical operating modes, i.e. build-up, capture and hold, during operation. Furthermore, the increased current needs in the vehicle must be secured and evaluated in an energy balance.

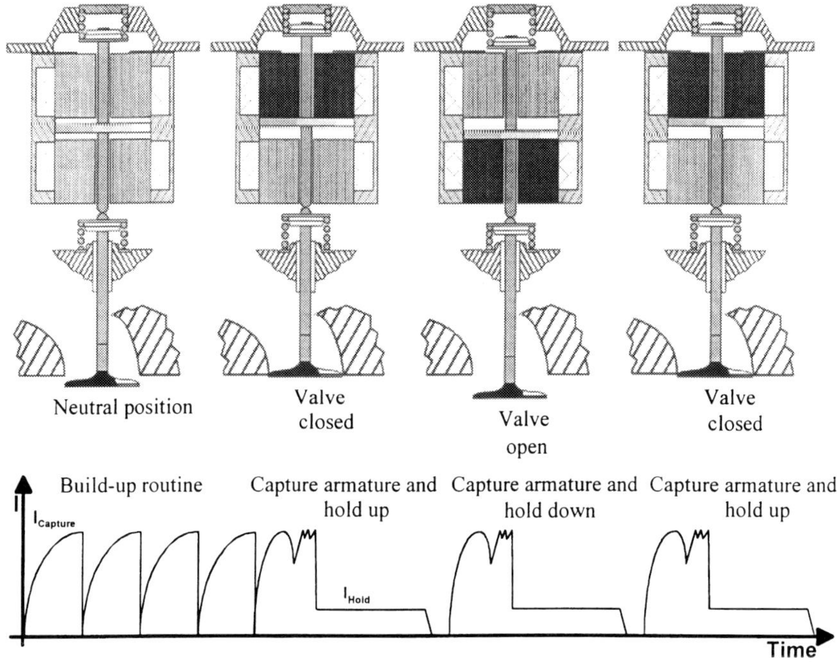

Fig. 23:   Mode of operation of electromagnetic valve control

5.3   Electrohydraulic variable valve control

The electrohydraulic valve control works in principle similar to the electromagnetic systems, as a free oscillator around the middle position, with the difference that the energy in the closed valve position is stored in the hydraulic spring and in the open position in the steel spring. By means of a 3/2-way valve, the electrohydraulic actuator is triggered according to the engine's needs and hydraulically deactivated through the control pistons in the end position. The system is in a de-energised state in the closed position, i.e. the exhaust and inlet valves are closed.

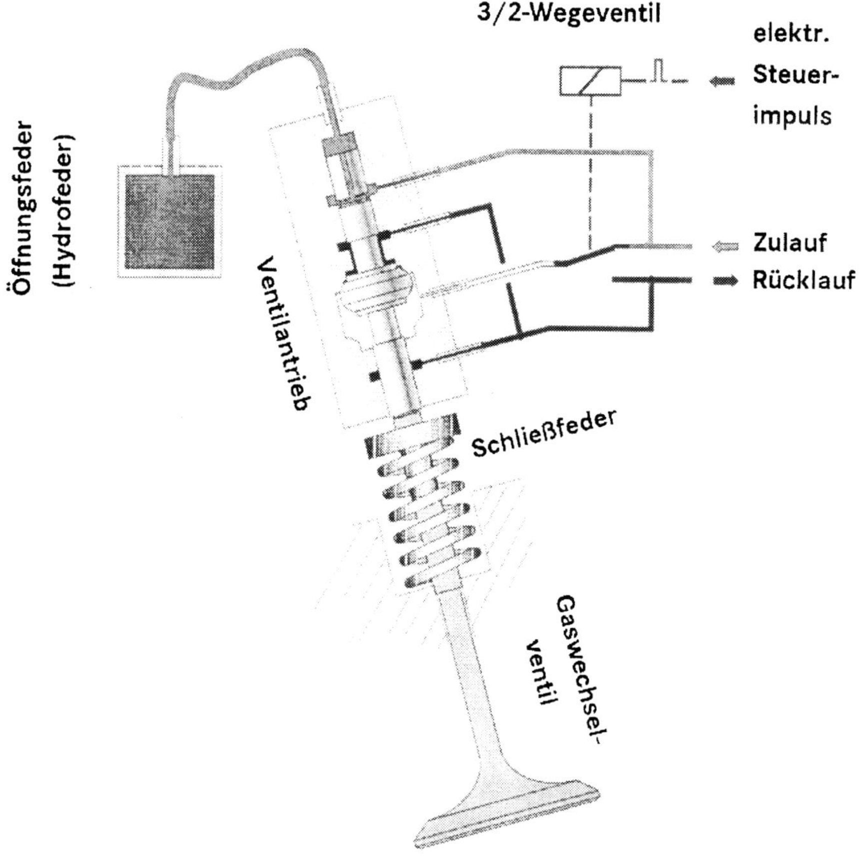

Fig. 24:   Electrohydraulic valve control

6 System comparison

The special features of the variable valve control concept are summarised in comparison to the conventional aspirating engine.

Under partial load operation with throttle-free load control a small portion of the reduced gas exchange requirement is consumed by the poorer high pressure efficiency. Nevertheless, an improvement in the effective efficiency of approx. 9% remains for the operating point examined, 2000 rpm and 2 bar pme. Attempts at raising the mean gas temperature with an optimised internal exhaust gas recirculation system and increasing the gas movement through installation of a seat tumble that is effective with small valve opening cross-sections are promising and resulted in an improvement in the high pressure efficiency. In the full load range medium-pressure increases can be achieved in the lower to medium speed range through free specification of the inlet valve closing time.

Criterion	Basic aspirating engine	VLC
Combustion concept	$\lambda=1$	$\lambda=1$
Mix formation	manifold	valve gap
Exhaust gas after-treatment	provided	provided
Variability of the control times	O	+
Fuel savings	O	+
System costs in comparison to basic unit (incl. exhaust gas after-treatment)	O	higher
Partial load		
- Pumping losses	O	+ +
- High pressure efficiency	O	–
- Gas temperature at BDC	O	–
- Final compression temperature	O	––
- Exhaust gas temperature	O	O(+)
Full load		
- Pumping losses	O	O
- Air requirement/filling	O	+
- High pressure efficiency	O	O

Fig. 25 Concept comparison

Large-volume engines with variable valve control achieve the greatest savings effects with a high partial load component in the overall effective load. Fig. 26 shows the gas exchange requirement $|\text{IMEP\_LPI}|$ for a conventional throttle-controlled aspirating engine and the range that is relevant for the test in accordance with the new European driving cycle (NEFZ) in a speed / medium pressure engine characteristics map.

The measurable impacts on fuel consumption depend substantially on the overall effective load used as the basis. Besides the reduced gas exchange losses, an additional filling improvement of up to 25% is possible, particularly in the lower speed range, through the variability of valve control. This has a very positive effect on the torque development and due to the reduced engine speed may implicitly lead to further fuel savings with equal operating performance through an extended rear axle.

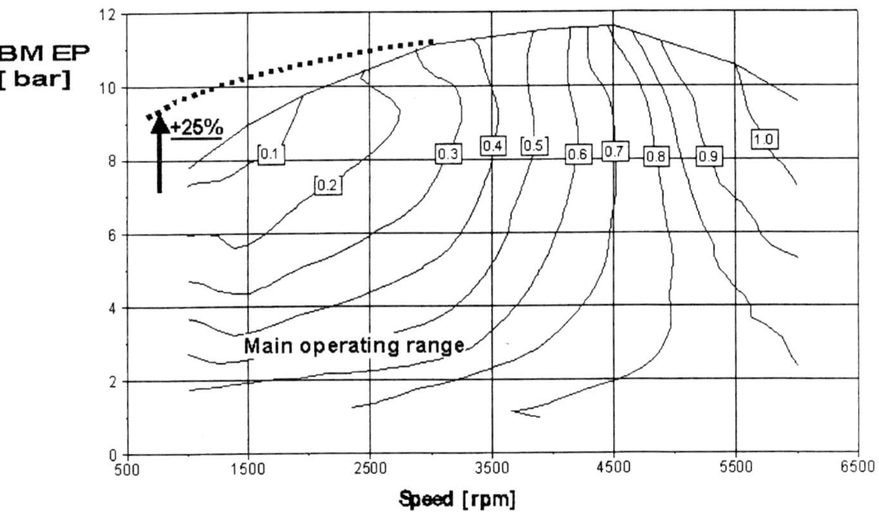

Fig. 26: IMEP_LP engine characteristics map

7 Summary

Throttle-free load control by means of variable valve control offers a promising way of improving, in particular, engines currently in production at a calculable expense and development risk.

Even in view of the currently intensively discussed concept of the direct injection internal combustion engine /3/, /4/, it appears meaningful to undertake parallel further development of variable valve control concepts, even considering the higher demands placed on exhaust gas purification technology of lean engines.

8 References

[1] F. Klein, M. Kühn, H.-J. Weimann, F. Wolpers, M. Krämer, J.Abthoff
*The influence of the valve stroke design in variable valve timing systems on load
cycle, mixture formation and the combustion process in conjunction with throttle-
free load governing.*
SAE Paper 98P-284

[2] F. Pischinger, T. Esch, T. Göbel, H. Richter
*Untersuchung der Gemischbildung bei unterschiedlichen
Lastregelungsverfahren für Ottomotoren*
Final report on FVV project no. 479 (AIF No. 8265)

[3] Dr.-Ing. G. Karl, Dr.-Ing. J. Abthoff, Dr.-Ing. M. Bargende,
Dipl.-Ing. R. Kemmler, Dipl.-Ing. M. Kühn, G. Bubeck,
Thermodynamische Analyse eines direkteinspritzenden Ottomotors
1996 Vienna Engine Symposium

[4] A. Frommelt, M. Bargende, M. Schulze
*Thermodynamischer Systemvergleich eines DE-Ottomotors mit konventionellen
Saugmotoren und Saugmotoren mit variabler Ventilsteuerung*
„Direkteinspritzung" Haus der Technik 1998

Authors' Index